FOOD FIGHTS

FOOD FIGHTS

How History Matters to Contemporary Food Debates

Edited by CHARLES C. LUDINGTON
and MATTHEW MORSE BOOKER

THE UNIVERSITY OF NORTH CAROLINA PRESS
Chapel Hill

Designed by Jamison Cockerham
Set in Arno, Clarendon, Scala Sans, and Avenir Next
by codeMantra

Cover illustration by Skillet Gilmore.

Manufactured in the United States of America

The University of North Carolina Press has been a member
of the Green Press Initiative since 2003.

LIBRARY OF CONGRESS CATALOGING-IN-PUBLICATION DATA
Names: Ludington, Charles, editor. | Booker, Matthew Morse, 1968– editor.
Title: Food fights : how history matters to contemporary food debates /
edited by Charles C. Ludington and Matthew Morse Booker.
Description: Chapel Hill : The University of North Carolina Press,
[2019] | Includes bibliographical references and index.
Identifiers: LCCN 2019019165| ISBN 9781469652887 (cloth : alk. paper) |
ISBN 9781469652894 (pbk : alk. paper) | ISBN 9781469652900 (ebook)
Subjects: LCSH: Food—United States. | Food—Biotechnology. |
Food—Safety measures. | LCGFT: Essays.
Classification: LCC TX360.U6 F675 2019 | DDC 363.19/20973—dc23
LC record available at https://lccn.loc.gov/2019019165

Contents

Figures, Tables, and Graph

FIGURES

TABLES

GRAPH

FOOD FIGHTS

Introduction

CHARLES C. LUDINGTON AND
MATTHEW MORSE BOOKER

In the United States today, it is commonplace, especially among the college-educated, to be interested in food and drink. This interest extends far beyond the gustatory and aesthetic questions of what to eat and drink, how to prepare it, or how to match the food with the drink. Instead, what we eat and drink, how we do so, when we do so, with whom we do so, where our food and drink is from, how it was produced, and the environmental impact, are considered by many Americans to be moral concerns, and thus political questions to be debated and then acted upon. It is no surprise therefore, that food is a subject of frequent and sometimes heated debate. In many of these discussions, however, one particular ingredient is often missing: history.

Surprising as it may seem to those who are so passionate about food, our contemporary food debates are not new. Rather, they have a history that reveals both continuity and change and a history that allows us to understand how and why we have the debates we do today. Consequently, this book is a collection of essays concerning some of the most common contemporary debates about food, in which the authors argue for their beliefs while also bringing historical context to the debate. We the editors, as well as all of the contributors to this volume, believe that the past matters immensely, because to understand the history of a debate is to begin to grasp the complexity of the issues being debated and the legitimacy of the debate itself. There are no simple solutions to today's food problems, but an awareness that today's problems grew out of yesterday's solutions, and that today's solutions may contribute to tomorrow's problems, can bring greater perspective and humility to everyone who cares about food both now and in the future.

Food Fights is premised on the importance of historical perspective, but it is also a book of essays written by scholars who believe that food, including beverages, is a core around which very different discussions in our society revolve. Readers of *Farm Journal* rarely overlap with readers of *Gastronomica*. Nor do viewers of Doctor Oz have much to say about *Chopped*, and neither seems to pay much attention to the Centers for Disease Control or the U.S. Food and Drug Administration. For all of these people, though, food is a central concern. The problem is clear: many Americans obsess about food because it is freighted with the hopes, fears, and anxieties of modern life. Yet discussions about food seem to operate as separate conversations, all valuable but rarely intersecting. The essays in this book are meant to create just such an intersection by joining popular food-related debate, in which history is so often overlooked, to scholarly food-related debate, in which history is so important.

Consequently, readers of this book will not always find straightforward commendation or condemnation of our current "food system" (the entire food chain from production to consumption to waste management). While some of the authors do take strong positions, all of them explore the myriad themes and tensions that define our current food fights. These include promises and failures of agricultural technology, winners and losers of industrial agriculture, and its sustainability. Taste, both for and of foods, how it is determined, and what it means, is another important theme of the book. So too is the issue responsibility for food supply and safety. Gender and cultural expectations, especially of women, receive significant attention, as does the meaning of cooking and eating. Finally, all of these essays about history are themselves products of particular historical viewpoints that reflect the long-standing nature of our contemporary food debates.

Food Fights grows out of a conference of food studies scholars organized by the editors at North Carolina State University. While historically grounded, the conference was purposely multidisciplinary. The goal of the conference was to move "from theory to practice," that is, to take stock of food history as a field of study and to propose new directions for research and writing.[1] For us, the conference organizers, what emerged was the need for a book that addressed the history of contemporary food debates and that explained to readers why the past matters to anyone engaged in those debates.

The past twenty-five years have seen tremendous growth in the field of food studies, and along with that growth has come the publication of myriad monographs, survey studies, and edited volumes of foundational texts, recent research, and state-of-the-field essays. This existing scholarship in food studies emphasizes a variety of disciplinary approaches as well as the numerous different domains into which food studies have been planted and born fruit; in many instances these works reveal how successfully food studies scholars have practiced the interdisciplinarity that is so often preached but rarely achieved in academic scholarship. This happy result is particularly clear in the work of food historians, who have combined the insights and methods of history with those of anthropology, sociology, psychology, economics, chemistry, biology, and medicine. In the process, food historians have demonstrated the nearly omnivorous nature of our species, the geographical, political, religious, social, and cultural reasons for what we eat and drink (and do not eat and drink), and the multitude of changes in the human diet from the distant past to the present.

While the list of recent state-of-the-field books in food studies is impressive, none of the currently available books emphasize the historical background and continuity of contemporary popular debates surrounding food. *Food Fights* fills that gap, and is a significant text in the growing corpus of food studies scholarship in general and food history scholarship in particular. Perhaps due to our position as scholars in a public, land grant, science and technology-oriented university, we are accustomed to arguing for and showing the importance of history and social context to current debates and events. We believe our academic context is a strength to be exploited and our book insists that history is not only fascinating, but that it matters in the present. *Food Fights* is therefore meant to provide historical context for contemporary debates, and to provoke debate. We have chosen the title *Food Fights* precisely because the chapters in our book illuminate the background of contemporary food debates but also partake in them. To be sure, while some contributors are more strident than others, none shies away from entering the debate she or he is addressing. This is fundamentally a book of history and not of policy proposals. Nonetheless, those who are looking for specific proposals will find much to chew on.

The primary goal of *Food Fights* is to bring a critical historical eye to food studies. This is important because while scholars in the natural sciences who work on food generally see themselves as practical problem solvers who have slowly but surely helped to increase crop yield, extend shelf life, and thus reduce starvation—and in wealthy countries to provide consumers with all sorts of specialty foods and drinks—scholars in the humanities and social sciences have overwhelmingly come to food studies from the perspective that the current American food system is unhealthy in almost every way. It is purportedly bad for the workers who produce our food, for the animals we eat (even before we slaughter them), for the land we farm, and the streams, rivers, and seas we exploit for irrigation, sanitation, and the food they contain. And of course the end result is that the entire system is bad for us, nutritionally, socially, and morally. This idea, which has informed and been informed by the popular food writing of such authors as Eric Schlosser, Michael Pollan, and Mark Bittman, has had two important consequences. First, criticism of the current food system implicitly condemns the eating habits of middle-class and poor Americans (and thus most Americans), the very people whom critics of the food system want to help. Yet telling people that they need to change their habits, even when accompanied by the caveat that such habits are not their own fault, is rarely a recipe for success.

The second problem of the unrelentingly negative view of our current food system is that it ignores what the system does well. Indeed, many of today's problems were the unintended consequences of policies and decisions that solved serious problems of a previous era. For example, critics pilloried the 2014 Farm Bill for disproportionately rewarding agribusiness over small farmers, for subsidizing wheat, soy, and corn, thereby discouraging the production of other crops and perpetuating American's addiction to high-fructose corn syrup (which contributes to America's world-beating levels of obesity-related diseases). All of this is true. Yet few of these critics knew or acknowledged that the original Farm Bill of 1933, from which all other farm bills have grown, was passed amid a global economic crisis by a progressive president and Congress who understandably wanted to guarantee that farmers could stay in business so that the United States would not have a starvation crisis on top of rampant unemployment. The point is not meant to gloss over or dismiss our current food-related problems in the United States; instead, it is to accept there are some very good reasons for the food system we have today. To ignore this fact is to risk throwing out the good with the bad.

We, the editors, certainly agree with many of the critiques of the current food system. We believe that our current food system could and should be improved for producers in particular, but also for consumers, as well as for animals and the environment. We believe that much more could be done by businesses, governments, and individuals to address the often-poor conditions and wages of food workers, the scourge of hunger and undernutrition, and the rapidly expanding problem of calorie-rich and nutrient-poor eating, and its attendant diseases. We think it's unfortunate that many people do not have the time, opportunity, knowledge, or desire to cook and eat regularly with family and friends. Humans are communal creatures, and working together to produce food, and then eating that food together (and sharing it with others), can be among the most fulfilling activities we do. Not surprisingly, eating communally has been shown to be good for our overall well-being, including that most crucial American obsession, living a longer life.[2]

But despite its faults, we also believe that there are some bright spots in the current food system and that all is not rotten to the core. Instead, we believe that our current food system is the creation of powerful historical forces. These forces are not naturally occurring but shaped by human agency more than anything else, and much of our current food system is the consequence of past efforts not just to make a profit, but also to solve problems of labor, cost, time, supply, demand, and nutrition. Thus, when we think historically, there is much to appreciate and enjoy in our current food system, just as there is much to criticize and improve on.

It should also be acknowledged that we the editors are "foodies" in that we think about food and drink from an intellectual perspective and an aesthetic perspective; we talk and write about the history of food and drink, and what it has meant in past societies, we love cooking and eating, we both garden, and we even grow our own oysters together. But we're also glad we're able to have day jobs as historians and still have time to cook in the evenings because someone else has farmed, raised, slaughtered, smoked, baked, fermented, packaged, and transported so much of our food for us. We appreciate that almost everything we eat is safe because of government oversight (and when safety is not guaranteed, we've made the choice). We like Whole Foods and specialty food stores, but we also shop at Food Lion and Walmart, and most definitely at Costco. We might prefer a grass-fed beef burger cooked rare and topped with a slice of red onion, in-season heirloom tomato, and spicy English mustard on a fresh, soft bun, but two or three times a year you will find us at the drive-through window at Burger King ordering a Double Whopper with cheese. And while it's impossible to beat the taste of deep-fried chunks of pork belly sprinkled with sea salt, the super-salty, fatty, crunchy pleasure of Fritos corn chips is not to be denied.

The chapters in *Food Fights* do not constitute a unified argument for or against the current food system; rather, we have purposely invited authors who hold very different perspectives on the current food system. Our book is about the history of contemporary food fights, and is itself a series of arguments about food. We strongly believe that the current food movement—in both its scholarly and consumerist form—is old enough, mature enough, serious enough, and powerful enough to deserve this sort of critical attention. Thus, just as Bill Cronon's edited volume *Uncommon Ground: Toward Reinventing Nature* (1995) historicized and scrutinized the

American environmental movement, *Food Fights* adopts a historical and critical perspective toward our contemporary food debates. Its fundamental purpose is not to assert that critics of our food system and popular diet are wrong and agribusiness and fast food binging are right. Nor is it to argue the opposite. Instead, *Food Fights* is meant to historicize our current debates about food in order to enhance popular appreciation of the strengths and weaknesses of our current food system and to make people more aware of what they are really arguing about when they argue about food.

We believe that scholars who condemn the current food system without acknowledging its successes unwittingly help to promote nostalgia for a golden age of cooking and eating that never existed. The medieval European peasants' imaginary land of Cockayne, where one could eat, drink, and then sleep all day, was only a dream. There is no reality there for us moderns to return to. Making food has always required someone's (usually arduous and time consuming) labor. Eating a diverse diet of tasty food has historically been the preserve of the few, not the many. Yet the myth of a golden era when people ate "authentic" food is a truism of the current consumerist foodie movement. In his book *In Defense of Food: An Eater's Manifesto* (2008), Michael Pollan specifically instructs his readers not to "eat anything your great-grandmother wouldn't recognize as food." Pollan was apparently unaware of the history of American eating in the period from the end of the Civil War until the 1980s, when so many American mothers (who became grandmothers and great-grandmothers) joyfully embraced the ease, convenience, and safety of packaged, canned, and powdered foods. Pollan and other critics of the current food system want us to eat locally, sustainably, and sparingly. That's certainly a laudable goal, as it promotes local economies, seasonal foods, and generally healthier bodies, but aside from getting the history all wrong, there are powerful historical reasons why so many Americans eat processed food from a global food chain that begins with the use of nitrogen fertilizer, and those reasons have much to do with the fact that producing food, from field to fork, is hard work that most people do not want to do. Furthermore, it remains doubtful that industrialized nations will revert to their former agrarian selves. Even those people who are most adamant about knowing where their food comes from are probably unwilling to work on a farm, a fishing trawler, or a slaughterhouse. And therein lies the conundrum of the current food movement. Without producing our own food, how do we create a food system and eating habits (including the joy of eating together) that are healthier for humans and the animals we eat and less damaging to the natural environment?

But asking such a difficult question does not settle the argument in favor of the status quo. Indeed, those who argue that our current food system works because it efficiently serves the interests of consumers are largely blind to the conditions of laborers who are needed to produce and distribute the world's food. Moreover, the environmental costs of our current food system are staggering. And there is no guarantee that scientists will find a solution to every problem that has been created. Supporters of our current food system also overlook the fact that choice, the alleged great good of consumer capitalism, is increasingly imaginary as one moves down the socioeconomic ladder. For most people cost, time, knowledge, cultural expectations and simple availability, restrict what food can actually be made, purchased and consumed. Moreover, the argument that our current food system, whatever its flaws, is the best of all possible systems rests upon the delusion that markets are merely an expression of human desires and not the constructions of unequal and often deeply unjust human relationships. So while *Food Fights* implicitly argues against those who would romanticize the past, it also argues against those who see our current food system as fair-as-fair-can-be given the conflicting needs and interests of government and business, producers and consumers. Nor would we, the editors, argue that a genuinely free market in food and drink could solve all of our food-related problems. Indeed, food history reveals perhaps better than any other form of history, the dangers of both over-regulated and under-regulated markets. Over-regulated markets skew prices, incentivize middlemen who leech the system, and create a surplus of some foods and a dearth of others; under-regulated markets invite unscrupulous hucksters of all sorts, create health hazards, and more importantly, go boom and bust. And nothing creates chaos and violence, or topples governments like a lack of staple foods. Thus, the issue that we face is not whether to regulate food markets; rather, it is what regulations serve society best. And that, of course, is the stuff of politics and the very thing in which the essays in this book engage.

The chapters in this book are organized to speak to each other and form five related sections, although it should be clear that each chapter speaks to many others, and thus the chapters should not be read only as a series of one-on-one combats. Section I focuses on perhaps the most intense food fight, regarding the benefits and banes of scientific innovation in food production in the field, laboratory, factory, and restaurant (or at least some restaurants). In chapter 1, Margaret Mellon addresses the hotly contested debates surrounding disputes over genetically modified crops. She writes from her perspective as a scientist to argue that genetically modified crops are safe but that genetic engineering has not delivered even half of what proponents promised. Agricultural genetic engineering solved some minor problems but in the process created many new ones. Her recommendation is to push forward with genetic engineering, but even more so with traditional methods of increasing plant yield and reducing blight, a call that is unpopular among corporations and scientific funding agencies. In the second chapter, Peter Coclanis takes a very different perspective while addressing the question of whether industrial agriculture (a.k.a. "Big Ag") is good or bad for American consumers. Coclanis argues that Big Ag is one of the great American success stories, that its industrial nature is the reason for its success, that corresponding environmental problems are small, and that science will solve those problems just as it solved other food production problems of the past. For Coclanis, it's full steam ahead. In chapter 3, Steve Striffler emphatically rejects Coclanis's favorable assessment of the current food system, not so much because of its environmental costs as because of its human costs, and in particular the costs to laborers. He argues that the two most common forms of contemporary food activism are both consumer driven. As a result, they both fail to address the greatest inequalities in our food system, which are seen among food laborers. It is only when we begin to focus on how we treat the people who grow and pick our crops, slaughter and dress our livestock, prepare our packaged meals, and work in the unseen parts of our restaurants, that we will begin to build a successful food system. Until then, argues Striffler, our food system works for some, but only because it occludes and oppresses others.

Section II addresses the different, although not entirely unrelated debate surrounding the role of food in establishing, maintaining and reflecting social hierarchy. In chapter 4, Margot Finn addresses the question of "taste" and whether the food we prefer as well as our foodways (how and what we prepare and eat) can be separated from social class. Finn looks at both theory and specific historical examples to conclude that taste is intrinsically bound to class, and that all food movements are ultimately class-based expressions of power or impotence. From this perspective, shopping at your local farmers market is less a moral act than a social act. Addressing a similar question in chapter 5, Charles Ludington asks why we tend to say "There is no accounting for taste" but then constantly make judgments about the things people eat. He mostly agrees with Finn that class background, cultural capital, and social aspirations determines what we as individuals find best. However, he argues that other concepts such as family customs (i.e., what we're used to), and various forms of what Ludington calls "tribal identity," as well as the desire to be perceived as authentic and the gendered meaning of the things we eat and drink, also influence the way we rank taste. More controversially, Ludington argues that some things we eat and drink are objectively better than others. To prove his points, Ludington examines the taste for wine in early modern Britain, and the taste for beer in recent American history. In a similar vein, Charlotte Biltekoff argues in chapter 6 that what American consumers imagine to be objective nutrition science has always in fact been driven by social values. In other words, nutrition science has been less scientific than most Americans imagine. Instead, what has driven the conclusions about proper nutrition has had much more to do with upper-class concerns about bodily and emotional control, and conceptions of beauty. As a result, nutrition advice is tailored not to individual needs but to "societal" (but actually elite) demands and expectations of what constitutes a healthy body, and what has come to be called "lifestyle." Health is thus "health," and it is no wonder that the nutrition advice we receive changes so often, and will continue to change.

Section III turns to the controversial subject of the government versus the market's role in food production. In chapter 7, Matthew Booker considers food safety legislation and the debates surrounding it in the United States. Booker examines the origins of these laws to show that Congress passed the 1906 Pure Food and Drug Act and companion Meat Inspection Act in an atmosphere rife with panic over disease and anxiety about the unlimited power of large, impersonal corporations. Focusing on deadly outbreaks linked to shellfish, Booker shows how everyday people, state and federal authorities, courts, and corporations sought to place responsibility for safe food on the government, a role that many businesses and libertarian minded Americans consider to be onerous. Likewise in chapter 8, Sarah Ludington examines the origins and debates surrounding government food subsidies. She looks at the history of the U.S. Department of Agriculture in the nineteenth century, and of the Farm Bill since its beginning in 1933, to show how that and subsequent farm bills illustrate the law of unintended consequences, from massive corporate farms driving small farms out of business, to the current obesity epidemic. Moreover, the history of the farm bills shows that government and big business are not nearly as oppositional as they are often made out to be by those on the political right and left. Instead, government legislation helps to create very powerful businesses and business lobbies, which in turn fashion government legislation in their favor, which eventually leads to demands for greater government oversight.

Section IV shines light on what was once an extremely vocal debate that is now sometimes hidden, and that is the question of women's roles in food production and serving, starting with the very first food humans eat, breast milk. In chapter 9, Amy Bentley looks at the long-standing controversies surrounding breast-feeding, formula, and baby food, to show that no lasting consensus has ever been established. This is in part because of changing advice from nutritionists (as per Biltekoff), but also because the entire debate reflects concerns about motherhood and the proper role of mothers in raising their children, which is itself an unsettled and probably perpetual question. Most notably Bentley's chapter shows that while women are still expected to raise their babies to become healthy children, they do so within rather stricter confines than we tend to imagine in our allegedly free and individualistic society. In chapter 10, Tracey Deutsch continues on the topic of women and food in order to address what role women should have in the kitchen. That may seem an odd question to many readers of this book, who are likely to respond, "Whatever role they want." But that has not historically been the answer. Nor is it even the answer that many Americans would give today. Deutsch shows how critics on both the right and left argue that women have let us all down by leaving the kitchen. Respecting the fact that cooking is hard work and many people do not enjoy it, Deutsch imagines a world in which men are more involved, and one that does not necessarily entail cooking solely within the context of a nuclear family.

Section V focuses on the contentious topic food and happiness. In chapter 11, Robert Valgenti addresses the question of whether caring about food actually matters for our quality of life. He examines philosophers' answers to this question since Plato, to show that food has rarely been considered essential to the good life, and at times it has even been seen as spiritually corrupting. Modern philosophers have been only slightly more charitable to food, but Valgenti shows that with the mind-body dichotomy increasingly rejected, food is coming into fashion among philosophers, and that's a good thing. Indeed, as Valgenti concludes, having a philosophy of food is integral to the good life. In chapter 12, Ken Albala wholeheartedly agrees with Valgenti by making an impassioned plea for the importance of cooking and eating with others for the sake of personal happiness. While Albala has written numerous books on food history from the Renaissance to the present, his contribution here is more personal than historical. He acknowledges that cooking and caring about eating well takes effort but that it is a form of work that connects us to the past, to farmers, to animals, and to our friends. In short, caring about cooking gives our food meaning. Moreover, it invites admiration from others, and all of these things together make those who cook happier human beings.

Finally, in chapter 13, Rachel Laudan addresses both Valgenti and Albala, while also returning the book to some of the debates raised in the opening chapters. Laudan's now-famous essay, first published in 2001, argues against a romantic view of food-related labor, whether in the field or the kitchen. She embraces processed and fast food, especially if it's made with high-quality ingredients. But in this volume Laudan adds a postscript that revisits her essay and articulates four different historical culinary philosophies in which she says we all partake, whether wittingly or not. Laudan thus categorizes the ways in which people in the West have tended to think about food production, preparation, and consumption. In so doing, she enters the fray by contextualizing all of us, including herself.

Laudan's postscript takes the broadest view of our food-related debates, and therefore provides a fitting coda to the book as a whole. However, hers is not the final word. This is, after all, a book that shows that our debates about food did not crop up overnight and will continue to be around in some form or another long into the future. Yet we are hopeful. While there are no easy solutions to our societal food fights, we believe that an awareness of the past will help all of us to create a better world in the future.

NOTES

1. "Food Fights: From Theory to Practice," North Carolina State University, Raleigh, 4–5 May 2012.

2. Cody C. Delistraty, "The Importance of Eating Together," *Atlantic Monthly*, 18 July 2014, https://www.theatlantic.com/health/archive/2014/07/the-importance-of-eating-together/374256/.

Section I
Producing Food

Section I addresses one of the most contentious food fights in the United States today. This debate is usually framed in terms of the alleged advantages and disadvantages of a heavily corporate, technology-driven food system. Namely, who benefits and who loses when food is more a product of agronomy than agriculture? One side generally argues that huge economies of scale and the use of the latest technologies has decreased prices, and increased the quality, quantity, and safety of the world's food supply, and will continue to do so. The other side agrees that food is relatively inexpensive in terms of the family budget and that quantity has increased, but they argue that quality has not kept up, that human health is imperiled by genetic engineering and toxic chemicals, that chemical fertilizers used in so much of modern farming are ruining ecosystems for generations to come, that increased quantity carries its own risk of excess, and, finally, that in the drive to increase profit and production alike, workers suffer. In other words, the vast majority of human beings are better fed than ever before, but our current system is both unjust and will destroy us in the long run.

Clearly, this debate matters because on the face of it the outcome will determine whether we are able to feed the planet or whether we return to starvation crises that have plagued so much of human history. Proponents of chemically based farming say theirs is the best and only way to feed the world's growing population, and they point to the successes of the "Green Revolution" of the second half of the twentieth century, in which chemical fertilizers, irrigation, and hybrid, high-yielding crops were promoted by international development agencies throughout the world. Opponents assert that the abundance we have today (in much of the world) will eventually be followed by dearth, high prices, disease, and war because of the deleterious effects of chemical fertilizers, insecticides, growth hormones, antibiotics, and genetic engineering on an ecosystem that has taken millions of years to develop. They point out that some of the things they've predicted have already happened. And of course they point to the often-dismal labor conditions that already exist.

The problem with the way this debate has been framed, is that both sides imagine Armageddon if they lose. That is possible, and it's a devil's wager to bet against either scenario. But perhaps a better way to frame the debate is by assuming that both these predictions are too bleak, and that instead we should be arguing over when and where to use the latest technologies to produce our food, when and where should we be using more traditional methods, how much should large corporations control our food supply, and, finally, how do we properly recognize and compensate the people who produce our food?

1

Savior or Monster?

The Truth about Genetically Engineered Agriculture

MARGARET MELLON

I first became aware of biotechnology early in the 1980s, when the field was in its infancy. Biotechnology arrived in Washington, D.C., on a wave of enthusiasm backed by the U.S. government, big corporations, and the scientific community. All of these entities had direct interests in the success of biotechnology: profits and influence for industry; global trade and economic clout for government; and grants and prestige for scientists. Citizens, too, were interested in the technology, not only in its potential benefits but also in its impacts on the environment and human health. But in those early days the optimism was high and the criticism was muted. During the last thirty years much of that initial euphoria surrounding biotechnology has waned. In particular, North Americans and Europeans are in the midst of a heated debate between the concentrated power of the direct stakeholders in the adoption of biotechnology, and a more diffuse set of stakeholders who are affected by the technology and who want a say in how it is used, or whether it should be used at all. Most of this debate over biotechnology is to be found in the domain of food, and the use of genetically engineered crops in particular.

AGRICULTURAL GENETIC ENGINEERING

Biotechnology is a broad term that can be applied to virtually any practical use of any living organism. Most uses—beer-brewing, beekeeping, and so on—are not controversial and many do not involve genetic modification of individual organisms. In this chapter, I will focus on techniques that are controversial because they involve modification of traits that can be passed onto subsequent generations and employ sophisticated molecular-level techniques to achieve modifications. Such techniques are referred to interchangeably as genetic engineering (GE) or genetic modification (GM). I prefer the term "genetic engineering" because it is narrower than "genetic modification" and more clearly excludes classical breeding and other time-tested methods of modifying organisms. Genetically engineered organisms that harbor combinations of traits that cannot be produced in nature are also sometimes referred to as transgenic. Genetic engineering can be applied to any organism—from a bacterium to a human—and the earliest commercial applications of genetic engineering, microorganisms modified to produce human pharmaceuticals, were not very controversial. However, this chapter will focus on genetic engineering of agricultural crops, a topic that has been, and for the foreseeable future will remain, hugely controversial.

MY INTRODUCTION TO BIOTECHNOLOGY

I was first introduced to biotechnology in the mid-1980s while working on the problem of environmental toxins at the Environmental Law Institute. Few in the environmental community could escape the excitement of the new technology that promised to transform industrial agriculture from a frequently toxic into an environmentally benign activity. As a scientist, I was naturally curious about the technology and predisposed to welcome it. Because of my background in molecular biology, many colleagues came to me with questions about it. I attended lectures and workshops on the issue and was invited by Monsanto to visit its headquarters in St. Louis, where I toured labs and greenhouses and heard the company lay out its vision for the new technology.

And a breathtaking vision it was: an agriculture that was no longer dependent on herbicides or insecticides; crops that could fix nitrogen and no longer need chemical fertilizer; crops that were innately high yielding; crops that could tolerate drought or cold; foods that could prevent disease; and agriculture so productive it would end world hunger. I heard the siren call. Without knowing much about agriculture, I expected that the technology would work, albeit with some unexpected or downside effects. I believed that regulation would help unearth and avoid such effects and was essential if the technology was to achieve its promise. I was not naive. I understood that regulation would not emerge without strong advocacy, but still I assumed that the technology would bring about big and benign changes. I decided to become a scientist-advocate in this field.

In 1992, I founded the National Biotechnology Policy Center at the National Wildlife Federation, an organization steeped in the mission of environmental conservation and protection, and committed to citizen activists as agents of change and government regulation as a way of facilitating input into important decisions. As I went into this work, I understood that industry, science, and citizens had different, but vital, roles to play in policy contests. I firmly rejected the idea that any players in this pageant were evil. My side and your side, for sure—but not villains and heroes.

At the policy center at the National Wildlife Federation, we accepted GE technology, regarding it as neither immoral nor uniquely dangerous. At the same time, we also rejected assertions that GE was inherently safe or necessarily better than alternatives. We felt free to accept some applications (drugs) while passing on others (many crops). We evaluated applications on three factors—risks, benefits, and the availability of alternatives. This position put the National Wildlife Federation in the middle of the advocacy spectrum—neither cheerleader nor adamantly opposed. Unlike those at the polar extremes, our middle position required data intensive evaluations of benefits, risks, and alternatives.

Benefits assessments quickly became the most challenging parts of our analyses. We came to understand that benefits depend on one's vision of agriculture. Herbicide-resistant crops, for example, were said to be beneficial because they encouraged farmers to use glyphosate, an herbicide less toxic than the more commonly used atrazine. To those who accepted that U.S. agriculture would continue to be structurally dependent on chemical pesticides, the replacement of one herbicide by a less toxic one counted as a benefit. However, because our goal was the minimization of chemical herbicides by using methods like cover crops and conservation tillage to control weeds, we believed that the substitution of one herbicides for another, without a commitment to overall herbicide use reduction, was not beneficial.

From my days as a lawyer, I was familiar with corporate influence and resources and understood that many public policy debates played out on an unequal field. But I was stunned by the clout that the preeminent biotechnology company, Monsanto, brought to the biotechnology debate. In order to promote its interests, Monsanto mounted major museum exhibitions, sponsored scholarships and fellowships at research universities, funded an entire wing at the Missouri Botanical Gardens, and more. It influenced domestic and international regulatory and political arenas, sometimes in disarming ways, and sometimes more aggressively. Despite its power, or perhaps because of it, Monsanto eventually lost the debate for the hearts and minds of consumers and others outside of mainstream agriculture, science, or government. In a 2015 Harris Poll of corporate reputations,[1] Monsanto ranked 97 out of 100 corporations. Recently, Monsanto admitted its problems[2,3] and initiated efforts to spruce up its image, including new approaches to social media.[4]

UNION OF CONCERNED SCIENTISTS

In 1993 I took a job at Union of Concerned Scientists (UCS), where I founded a program focused on agriculture rather than biotechnology. Because of the issues UCS took on, it was important for advocates at UCS to maintain scientific credibility. In that regard I was pleased to be named a fellow of the American Association for the Advancement of Science in 1994 and a Distinguished Alumnae of Purdue University in 1993, which established my scientific bona fides. My experience at UCS taught me about the role of science in big societal debates. UCS's signature issue was nuclear power plant safety. Questions on how to build and run a plant or what safety measures might work are purely scientific questions, the necessary bedrock of the debate. But the debate was about more than science. Questions of how much risk to take in exchange for the benefits of nuclear power, or whether nuclear power is preferable to coal or wind power, cannot be answered by science alone. They require societal judgments on which reasonable people can and do disagree. Both sides in the nuclear power debate make arguments based on science. But, a favorable view of nuclear power does not make one pro-science; nor do concerns about the safety of nuclear power make one anti-science.

In contrast, the biotechnology debate I became involved in at UCS has been cast as a debate between science and anti-science. Critics of genetically engineered crops are labeled Luddites and arguments against the technology are called emotional and sidelined as illegitimate. But the genetic engineering debate involves economic, health, and safety issues that are rife with societal judgments. Societal questions of how much risk is appropriate for the supposed benefits of the technology, for example, involve far more than scientific considerations. Not surprisingly, simplifying and polarizing the genetic engineering debate in this way has seriously distorted it.

One reason the genetic engineering debate is so intensely contentious is that it sits at the juncture of many overlapping issues: food safety; environmental safety; control of the food system; international trade; wariness of technology; deep-seated mistrust of government; and animal welfare. Science can be a component of all of these issues, and getting the science right is crucial to productive debates. But science per se does not have a position on environmental safety or control of the food system. Like nuclear power plant safety, the safety and appropriate use of GE crops raise societal issues on which reasonable people can and do disagree.

TENSIONS OF SCIENCE ADVOCACY

Scientific advocacy organizations like the National Wildlife Federation and the Union of Concerned Scientists (UCS) seek to change the world for the better, often through the legislative process. At NWF and UCS scientists work closely with lobbyists and media experts to carry out legislative campaigns. In so doing, they need to communicate with and motivate citizens who have much else to think about. One of the best ways to mobilizing citizens is with short, compelling messages. Such messages are often stylistically incompatible with the highly qualified language of science, which values precision and abhors overstatement. Crafting such messages—for action alerts, for example—routinely leads to lively discussions between scientists and media professionals in advocacy organizations. In the genetic engineering debate, some opponents of genetically engineered foods have mobilized supporters with scary images like skulls and crossbones, scarecrows, or evocative terms like "Frankenfood."

While I worked closely with the media professionals in my organizations to craft strong messages, I drew a line well short of using the term "Frankenfood" or Halloween imagery in communications with the public. I believed these images conveyed a degree of risk not warranted by the early products of the technology and that they preclude the practical and the fact-based debate we need about the impact of the technology in food and agriculture. However, I soon found that critics of genetically engineered crops are not the only ones to use short, evocative—and misleading—messages in these debates. Proponents do the same. The most egregious example is the campaign to convince the public that genetic engineering is necessary to "feed the world," a message delivered with images of smiling farmers from developing countries. The implication is that use of genetic engineering in the United States and elsewhere is essential to meeting the challenges of world hunger and that critics of genetic engineering are impeding the only solution to the problem. But scientists understand that the root causes of world hunger are complex and, for the most part, grow out of poverty, not out of challenges in agricultural productivity.[5] To suggest that genetic engineering by itself is the magic solution to world hunger is just as misleading as suggesting that it is inherently scary. On both ends of the spectrum, these communications strategies are a major challenge to nuanced, scientifically sound debate.

LOOKING AT BENEFITS AND ALTERNATIVES

The policy thrust of my early advocacy focused on the risks of products of genetic engineering and the need for government regulation to control those risks. The scientific analyses I and my collaborators produced in my early career, assessed the ecological risks of genetically engineered crops;[6,7] the threats to the efficacy of *Bacillus thuringiensis* (*Bt*)—a biological pesticide—posed by resistance;[8] gene flow of GE traits within agricultural crops,[9] and the uncontrolled movement of pharmaceutical traits in cultivated and wild environments.[10]

By the earlier 2000s, biotechnology's most popular crops had a track record that could be evaluated both against a sustainable vision of agriculture and the grand early promises of the proponents. In 2009 and 2011 my talented friend and colleague Doug Gurian-Sherman produced landmark studies asking what GE technology had accomplished in three key areas—yield,[11] nitrogen use efficiency,[12] and drought tolerance[13]—all of which were among the dramatic improvements promised by genetically engineered crops. In each study, Dr. Gurian-Sherman assessed the performance of alternatives to genetic engineering like classical breeding, agroecology, and an enhanced version of classical breeding called marker-assisted breeding. Gurian-Sherman's careful analyses of the benefits of GE crops led to a big surprise.

Despite wide adoption and commercial success, overall biotechnology has a disappointing track record. As an agricultural technology, it had not achieved even a modicum of what it had promised. Beyond that, Gurian-Sherman's analyses showed that in the very areas where biotechnology had stumbled, classical breeding and agroecology had succeeded. For me, as well as for many other scientists in the field, Doug's work was an eye-opener that prompted many questions. As an early enthusiast for the promises of genetically engineered crops, I had to face the reality of its disappointing performance. So, in midcareer, I refocused my interests on the articulation of alternative, more sustainable visions for agriculture and strategies for achieving them (e.g., in the 2013 Vision Statement of the Union of Concerned Scientists).[14] And it is to that subject which I will now turn.

CHEMICAL-FREE AGRICULTURE?

Let's start with perhaps the biggest promise of all: genetic engineering will allow us to achieve chemical free agriculture. Has this been achieved? Not even close, although there have been a few bright spots. There are no GE nitrogen-fixing crops,[15] and while the herbicide-resistant crops reduced herbicide use initially, the trend is now heading in exactly the opposite direction because weeds, like all organisms, can adapt to stress.

Among herbicide resistant crops, Roundup Ready™ crops are the commercial stars in the biotechnology pantheon. Companies have introduced commercially successful varieties of most of the important commodity crops: soybeans, corn, cotton, alfalfa, and canola. These crops were widely adopted because—at the beginning—they saved time and reduced costs, especially on large industrial farms. As a result, one herbicide—glyphosate—has been used on tens of millions of acres of American farmland, year after year. Predictably, such intensive use encouraged the growth of weeds able to withstand glyphosate, and soon such weeds began to show up in fields all over farm country.[16] Farmers responded by using more and more glyphosate and adding other chemical herbicides to the mix. Thus, the early dip in herbicide use was soon reversed and herbicide use skyrocketed. Over its first sixteen years, the evolution of resistant weeds led to a 527-million-pound *increase* in herbicide use in the United States.[17] Because farmers continue to plant the GE crops, and resistant weeds continue to emerge, herbicide use is still rising.

The industry's response has been to engineer resistance to additional herbicides, 2,4-D and dicamba, into crops, so those chemicals can be applied along with glyphosate.[18] In sum, widespread use of herbicide-resistant crops, by far the dominant application of agricultural biotechnology, has lead to an unfolding environmental and agronomic disaster: more and worse weeds, higher farm costs, exploding use of herbicides, and the evolution of multi-herbicide-resistant weeds.[19] These concerns are heightened by the International Agency for Research on Cancer's recent determination that glyphosate is a probable human carcinogen[20] and that 2,4-D, whose use will increase with the next generation of herbicide-resistant corn and soybeans, is a possible carcinogen.[21]

The second major application of genetic engineering has been crops modified to produce their own insecticidal toxins.[22] The toxins were originally found in soil microbes called *Bacillus thuringiensis* (*Bt*). The family of *Bt* toxins contains slightly different molecules that kill different classes of insect pests. Like herbicide-tolerance traits, *Bt* toxins have been engineered into a variety of crops, most importantly corn and cotton. Over the first sixteen years of the biotechnology era, insecticide use in this country was reduced by 123 million pounds.[23] Although the decline was offset by the dramatic rise in herbicide use, the environmental benefits of lower external insecticide use in *Bt* crops have been impressive.[24] And until recently the emergence of *Bt* resistance was held at bay. One big reason for this success has been the implementation of sophisticated refuge strategies developed by entomologists to delay resistance.

But trouble is brewing in *Bt* crop fields. Farmers did not heed entomologists' advice on a major pest of corn—the Western corn rootworm—and now fields are teeming with these rootworms despite being planted with *Bt* corn varieties.[25] Belatedly, the Environmental Protection Agency has developed plan for companies and farmers to manage resistance in the rootworm,[26] but it may be too late. In addition, the introduction of *Bt* crops has coincided with the increased use of other insecticides in corn systems, most ominously the neonicotinoids or "neonics." These chemicals, first introduced in the early 1990s, are now the most widely used insecticides on earth. Neonics are highly toxic to insects, including honeybees and other pollinators at very low doses, and their nearly ubiquitous use is a suspected cause of the decline of bee colonies around the world.

The rise in neonic use was missed initially because the pesticides were sold as seed coatings and as such were not counted by the government in surveys of pesticide use. While not a direct result of GE technology, the demand for neonics certainly belies the promise that biotech crops would usher in an era of chemical-free agriculture. In sum, while deserving credit for substantial reductions in insecticide use in corn and cotton, *Bt* crops have not staved off ever-increasing use chemical insecticides in those crops. Instead, fields full of genetically engineered crops are saturated with chemical poisons.[27,28]

Having said that, genetic engineering has had success with crops engineered to be resistant to viruses, but the total acreage of virus-resistant crops is tiny in comparison to herbicide-resistant and *Bt* crops. A virus-resistant variety of papaya has been widely adopted by Hawaiian papaya growers and continues to account for 70% of the Hawaiian papaya crop.[29] Varieties of virus-resistant squash have also been approved for sale in the United States, although there are no mechanisms for determining how much of the seed has been sold.

What then can be said about the dream of chemical free agriculture that was envisioned by proponents of genetically engineered crops some thirty years ago? Despite a few bright spots, large-scale adoption of GE crops has led to dramatic increases in pesticide use that are likely to continue to increase in the future. Put another way, however hopeful our dreams may have been, GE in practice has turned out to be a chemical and environmental nightmare.

HIGH-YIELDING CROPS

Another claim made in the salad days of biotech agriculture was that genetically engineered crops would produce much higher yields. In fact, that has not been the case. But first it's important to understand that crop yields are of two types. The first type of yield refers to performance in the presence of pests or stress. Herbicides and insecticides increase yields when weeds or insects are present, but have little effect when pests are absent; those are called operational yields. The second, more fundamental kind of yield is innate or potential yield, the yield possible under the ideal conditions—with no pests, no stress, adequate nutrients, and benign climate. Innate yields represent the upper limit on agricultural productivity, and increasing them is essential to keeping pace with increasing human populations.

Classical breeding is the stellar technology in this realm. It is responsible for virtually all the increases in innate yield in crops since the dawn of agriculture. Dramatic examples of the power of classical breeding are the shorter, sturdier versions of wheat and corn yields that were the backbone of the Green Revolution.[30] Less dramatic, but no less important, are the steady, ongoing 1–2% a year annual increases in U.S. corn yield that are attributable to classical breeding and agronomic practices. By contrast, as of 2015, there is only one new GE crop pending in the commercial pipeline that claims to increase the innate yield of a crop. This dismal performance on innate yields is often missed because of confusion with increases in operational yields—yields measured in the presence of pests—that have been produced by *Bt* crops.[31] Nevertheless, GE's failure to produce crops with increased innate yields severely undercuts the claim that GE is essential for fundamentally improving agriculture.

In a similar vein, scientists once hoped and predicted that crops could be genetically engineered to resist various forms of stress. Alas, genetic engineering has produced only one commercialized stress-resistant crop, Monsanto's DroughtGard™ Hybrids, corn varieties resistant to mild drought. No commercialized GE crops are resistant to being flooded (important in rice production), to cold, or to salt. Sometimes confused with drought tolerance, water use efficiency is the ability to maximize yields from a given amount of water, often irrigation water.[32] Again, there are no GE crops on the market in this category. Monsanto's public relations materials include posters asking, "How Can We Squeeze More Food from a Raindrop?" The answer is, if we rely on current GE technology to produce water-use-efficient crops, we can't.

FOODS AS MEDICINE

Of all the promises of genetically engineered crops, perhaps the most exciting was that they could be engineered to prevent human disease. In this regard, GE so far has been a disappointment, although not for lack of trying. Right now, there are no GE crops on the market engineered to prevent disease. There are some products that claim more nutritious oils and one that reduces acrylamide levels,[33] but there are no studies demonstrating health benefits from consumption of these foods. Perhaps most famous of all disease prevention crops is golden rice, a rice that has been genetically engineered to combat vitamin A deficiency and the nearly 700,000 annual childhood deaths that result from this deficiency each year. However, after almost twenty years of effort, golden rice has still not gotten the green light from its sponsor, the International Rice Research Institute (IRRI). According to IRRI, the yields of the rice still lag behind comparable varieties in Philippine fields.[34] Moreover, golden rice has yet to be shown to increase vitamin A levels in target populations, perhaps because the diets of extremely poor people lack sufficient fat to enable the absorption of the vitamin.

ENDING WORLD HUNGER?

Finally, what can we say about the claim that genetically engineered crops would improve agriculture in developing countries and alleviate, if not even eliminate, world hunger? There have been successes with genetically engineered crops in the developing world, primarily with a fiber crop, *Bt* cotton.

But the promise that the technology alone could produce necessary changes in the complex arenas of developing country agriculture and world hunger (a phenomenon not confined to the developing world) has not been fulfilled. The only large consensus international study of developing-country agriculture concluded that agroecological approaches have more potential than GE crops for meeting future food and agricultural challenges.[35] This conclusion should not be surprising. It is common sense that a crop-breeding technology whose major successes after twenty-five years are confined almost entirely to the realm of pest management cannot be counted on for the broad array of products needed to improve agriculture worldwide. It is possible that the situation will change in the future; genetic engineering may one day deliver a robust array of crop varieties with traits like increased innate yield, stress tolerance, and a host of other benefits. But right now the development pipeline in the United States is dominated by pest management products, primarily more combinations of *Bt* and herbicide-resistance genes. This is hardly what we once believed GE crops could do.

THE IMPLICATIONS OF GENETIC ENGINEERING AS A LIMITED TECHNOLOGY

The performance of GE in its first several decades has fallen short of original expectations, primarily because of technical challenges. So far the major applications of genetic engineering have been based on the transfer of one or a few independently operating genes. Herbicide tolerance and *Bt* toxin production are essentially one-gene traits. A number of the herbicide tolerance and *Bt* genes can be added successively to crops as so-called stacks, but the genes do not interact with one another. Yet many important crop traits like innate yield and stress tolerance require multiple interacting genes. Unable to move interacting sets of genes, genetic engineers, for the most part, have failed to improve such traits.

Another barrier to the success of genetically engineered crops is the imprecise nature of the processes for delivering new genetic material to crop plants. The most popular methods carry pieces of genetic material into plant cells and integrate them into chromosomes where they can function. But this method inserts the new genetic material more or less randomly along the length of chromosomes, and often at more than one location. The inability to control the number and location of gene insertions makes genetic engineering a hit-and-miss—and expensive—process.

SUMMARY OF BENEFITS OF GENETICALLY ENGINEERED CROPS AFTER TWENTY-PLUS YEARS

Although it has succeeded with a very limited set of pest control applications, agricultural biotechnology has produced substantial economic benefits for a handful of entities. Sales of herbicides, insecticides, and seeds have generated huge profits for technology companies. And until resistance sets in, farmers benefit economically from reduced input costs for weed and insect control and increased convenience, especially in large commodity crop operations. As mentioned above, availability of the virus-resistant GE papaya reinvigorated the Hawaii papaya industry that was previously threatened by the ring-spot virus.[36]

There have also been some public benefits stemming from the use of genetically engineered crops. By public benefits, I mean improvements in the quality of shared resources like air, water, soil, or health. Public benefits of *Bt* crops have included reduced insecticide use on the *Bt* crops, including so-called halo effects in nearby non-*Bt* crops, which can benefit indirectly from suppressed pest populations.[37] Adoption of herbicide-resistant crops also led to reduced herbicide use in the early days before the appearance of resistant weeds. Herbicide-resistant crops may encourage adoption of conservation tillage, including no-till, wherein a new crop is planted through the remnants of the previous year's crop. Conservation tillage helps to prevent water erosion and at one time was credited with sequestration of carbon in soil, but careful scientific studies have shown that that is not the case.[38,39] In any case, the majority of the increase in the use of conservation tillage predated the introduction of the GE crops.[40] In short, the public benefits of GE crops should not be dismissed, but they are hardly overwhelming.

CROP IMPROVEMENT ALTERNATIVES SUPERIOR TO GENETIC ENGINEERING

A key fact often lost sight of in the biotechnology debate is that genetic engineering is not the only way to improve agricultural crops and animals. Classical breeding; marker-assisted breeding, which is an enhanced version of classical breeding using molecular biology to identify parental organisms (also called marker assisted selection); and agroecology, a scientific discipline that uses ecological theory to design and manage agricultural systems that are productive and resource conserving, are all powerful alternatives to genetic engineering.[41] More importantly, in many cases these techniques have been scientifically proven to be superior to GE in improving agronomic traits across the breadth of crop varieties. In fact, most of what GE has claimed to have done, or is trying to do, has already been done using one of these approaches.[42,43,44] There are areas where classical breeding faces challenges, like the production of vitamin A precursors in rice (the goal of golden rice), but they are relatively few.

To cite a few examples of success, classical breeding routinely turns out crops with increased innate yields—indeed it is the only technology that ever has. Let me repeat: classical breeding is responsible for virtually all increases in innate yield since the beginning of agriculture. The same is true for crops with increased nitrogen use efficiency and water use efficiency—there are no GE varieties, but many classically bred varieties.[45,46] Even where GE does work, it is often slower and more expensive than classical breeding. A project launched by the International Maize and Wheat Improvement Center in 2012 was reported in 2014 to have developed twenty-one classically bred varieties of nitrogen-use efficient crops adapted to African soils, while the single comparable GE variety to come out of the project was at least ten years away.[47] Classical breeding has also produced a ring-spot resistant variety of papaya,[48] carotene-rich sweet potatoes,[49] and a nonbrowning apple.[50] Classical breeding, enhanced by marker assisted breeding, has also been successful in producing flood-resistant rice varieties now found in rice paddies all over Asia.[51]

AGROECOLOGY: CROP ROTATION COVER CROPS AND OTHER AGRONOMIC PRACTICES

Crop rotation, cover crops, and other agronomic practices help prevent the emergence of pests in ways that are superior to engineered crops aimed at only one or two pests. These practices exemplify an agroecological as opposed to an industrial approach to agriculture. Agroecology uses ecological science to design systems that minimize or prevent pest problems, while industrial systems usually address pests with toxic chemicals. Planting different crops in the same field in successive years deprives many potential pests of reliable food sources and prevents their buildup, reducing the need for pesticides from the get-go. Scientists concerned about the emergence of the resistant corn rootworms have noted that three- or four-crop rotations would make it unnecessary to use either *Bt* corn or chemical pesticides. One group of public interest entomologists commenting on the EPA resistance management plan for corn rootworm recommended that crop rotation be the sole course of action in response to evidence of resistant rootworms.[52] Unlike industrial systems, which are characterized by the evolution of resistance and the subsequent need to turn to new pesticides (the so-called pesticide treadmill), properly implemented agroecological approaches are robust over time. Insects have a hard time developing resistance to them, so they preserve rather than erode the efficacy of chemical pesticides.

WHEN FACTS DON'T MATTER

My biggest personal disappointment over the course of my career has been how stubbornly the biotechnology debate has resisted recognition of classical plant breeding and agroecology as plausible, if not preferable, alternatives to genetic engineering. In light of the solid science supporting the broad potential of such approaches, it seems that facts just don't matter. Over the years I have tried many times to interest the media in the performance of crops produced by classical breeding, but was always rebuffed. GE's vitamin A rice made it onto the cover of *Time* magazine on the strength of a mere whiff of potential to help hungry people with nutritional deficiencies. Meanwhile the hundreds of classically bred rice varieties produced by IRRI that are already in the ground, boosting health, nutrition, and farmer incomes, stir very little media interest. Similarly, the U.S. Department of Agriculture (USDA) seems unmoved by track record of classical breeding. My suggestion that the USDA produce a brochure touting the vital role classical breeding plays in U.S. agriculture was met with a big yawn.

Some of this reaction is certainly due to the power of commercial interests in modern agriculture. Industrial systems are dependent on chemical inputs and chemical companies, like Monsanto, have tremendous influence with media and governments. Producers of pesticides, whether placed inside or outside a crop, have little interest in pest management solutions that reduce pesticide use.

But I believe the preference for GE comes from a deeper place in our history. For many Americans, new technologies embody the idea of progress, a central idea of both the Enlightenment era in which our nation was born, and of the liberal capitalism that we have historically practiced. This conception of progress is rooted in the belief that scientific discoveries and technological advancements are necessarily superior to that which came before. Thus, we look forward to a constant stream of innovation to achieve better lives and more vibrant societies. Our faith in progress leads us to prefer new to existing technologies even when the new ones aren't working particularly well. We tend to give new technologies the benefit of the doubt. If they don't perform this year, they may do so next year. Faith in progress even overrides concerns about harm. To make omelets, we must break eggs.

Progress is one of the many mental frames that allow people to organize and respond to facts and arguments, and frames are notoriously resistant to facts. I believe the progress frame blocks the acceptance of the truth of the GE technology's relatively poor performance. But reality matters. Or at least it should. We gave transgenic technology a chance to prove itself, and it did not. That was a reasonable but costly miscalculation that has deprived the world of more effective technologies available to confront agricultural challenges.

So I happily acknowledge Peter Coclanis's point in this volume that industrial agriculture is highly productive and historically fueled the industrialization of the rest of the American economy. But the relentless drive toward productivity and the belief in technologically fueled progress has overshot its mark. Only 2% of the population now is engaged in commercial farming, yet the United States can feed itself many times over. The challenge in agriculture is no longer producing enough to feed ourselves, but what to do with our enormous agricultural overproduction. That's why we devote huge swaths of fertile midwestern farmland to energy crops, not food crops.

But the intense focus on productivity and the faith in new technologies have had other important consequences as well, including environmental problems that are growing in magnitude. Toxic pollution of air and water, loss of soil fertility, production of climate gases, loss of pollinators, and the disruption of the nitrogen cycle and the destruction of coastal ecosystems can no longer be dismissed as mere externalities to be managed or ignored. Taken together, they are becoming existential threats to the planet.

Innovators have developed new sustainable systems of agriculture that can dampen threats to the environment without undermining productivity but they require a fundamental shift away from our current system based on enormous monocultures of a handful of crops. That move will require bold new ideas: reintegration of plant and animal agriculture, multiyear crop rotations, reducing our consumption of meat, and much else.

It won't be easy. As Rachel Laudan lays out in her essay (chapter 13), agriculture and food are complicated issues—environmentally, culturally, and economically. Coclanis underscores how a willingness to change has been an important element in the past success of U.S. agriculture. This is certainly true. Unfortunately, however, rather than chart a course for change, most of the current agricultural establishment has battened down the hatches, defending the status quo by belittling creative entrepreneurs like organic farmers. The biggest obstacles to the improvement of agriculture today are the inability to admit its environmental shortcomings and the failure to envision fundamentally new ways of doing things.

JUST A TOOL IN THE TOOLBOX

Yet all is not lost. An accommodation to reality is under way. Many proponents of biotechnology have abandoned the early grand claims of a transformational technology and now modestly present GE as just a tool in the toolbox. The argument now is that because society faces momentous challenges, we need every available tool to solve them; so, although not a panacea, GE is still important. And I agree, we do need all the tools in the toolbox. But we need to take seriously GE's poor track record. Yes, GE is a tool, but the evidence shows it to be a limited one. Meanwhile, it is taking up most of the room in the box, pushing bigger and better tools to the side. Resources for agricultural research, extension, and trade promotion continue to go disproportionately to GE products.

If we are serious about taking on the challenges of burgeoning populations, environmental degradation, and climate change, we need to make room for the big tools: classical breeding, marker-assisted breeding, and agroecology—and fund them commensurate with their importance. And we need help from the media to talk about the big tools. How about a front-page story in the *New York Times* on the contribution of classically bred crops in developing-country agriculture?

To be clear, in my view, GE crops do have a role in the future of agriculture, but the role will be relatively small—primarily as niche products in the few areas where classical breeding falls short. Putting the cart of genetic engineering before the horse of sustainable systems has impeded society's ability to produce sufficient food in a warming, polluted world. Our greatest progress will be made not by reflexively adopting new technologies but by choosing wisely from among the technologies available to us, especially those with proven track records of success.

HERE WE GO AGAIN?

What then of the future of genetic engineering? I just said that I think GE should be a small tool in the agricultural toolbox, and I've tried to make it clear that the transgenic techniques that drove the early biotechnology revolution are beginning to run out of steam. The combination of the technical challenges and the paucity of new useful proteins is putting a damper on the era of transgenes.

However, moving new genes into crops is not the only GE technique. The next wave of genetic engineering will probably be based less on adding new genes and more on editing and modifying the genes already in place in the target crops. These techniques are emerging from exciting new discoveries about the elegant but complicated ways gene expression is controlled in cells.[53] Scientists have learned that noncoding stretches of DNA can give rise to a variety of RNA molecules that determine which genes produce proteins and when. Double-stranded RNAs are key molecules in these processes, which have given rise to two general categories of techniques—gene silencing and gene editing. Gene silencing, or RNA interference (also called RNAi), turns genes off by inactivating or degrading messenger RNA molecules that code for proteins.[54] Gene editing can snip into DNA and stimulate repair processes that introduce small or large changes in the DNA sequence at specific locations.[55]

The ability to produce changes at particular locations in DNA molecules offers a big advantage over techniques that insert new genetic material into chromosomes at random locations. Both editing and silencing have proven to be powerful research tools for deciphering the secrets of gene expression and scientists are working diligently to employ them in agriculture and human medicine. Gene editing and gene silencing, however, have downsides. They cannot restrict their snipping activities to the target site and often cause off-target effects.[56] These are significant problems.

Nevertheless, the future of these techniques looks to be exciting, and once again we are hearing the siren song of new technologies that will change everything.[57] This may well happen, and the possibility of new technologies able to address agricultural and health problems should be welcomed. But our experience with agricultural transgenesis should temper our enthusiasm. This time around, we can no longer afford to be blinded by the light of promises and possibilities. Instead, we need to invest resources to understand and assess the risks and benefits of these new techniques. We also need to ask questions early on about whom the technologies benefit and what the alternatives are. Even if gene silencing can control a single pest, is it a better approach to pest management than crop rotation, which can control many pests? If gene editing and silencing are directed to agricultural productivity, will they deliver innate yield or stress tolerance any better than classical breeding or agroecology do?

Dealing with new technologies is a central challenge of modern life. We need to find the balance between the hype and enthusiasm that fuels innovation and the hard-nosed evaluation of risks, benefits, and especially alternatives. So, while I don't call genetically engineered crops Frankenfood, neither do I call them the future.

NOTES

In 2014, at the invitation of the Genetic Engineering and Society Center, I had the privilege of delivering a public lecture at North Carolina State University exploring the history debate around biotechnology in the United States. I welcomed the opportunity to reflect on issues I have worked on for most of my career. They have turned out to be more fundamental and far-reaching than I ever imagined when I first encountered them. This essay is adapted from that lecture.

1. Harris Poll, "Regional Grocer Wegmans Unseats Amazon to Claim Top Corporate Reputation Ranking," 4 February 2015, http://www.prnewswire.com/news-releases/regional-grocer-wegmans-unseats-amazon-to-claim-top-corporate-reputation-ranking-300030637.html.

2. Jenny Hopkinson, "Monsanto Confronts Devilish Public Image Problem," *Politico*, 29 November 2013, n.p.

3. Lessley Anderson, "Why Does Everyone Hate Monsanto?," *Modern Farmer*, 4 March 2014, n.p.

4. Sarah Henry, "Monsanto Woos Mommy Bloggers," *Modern Farmer*, 18 September 2014, n.p.

5. Canadian Biotechnology Action Network, "Do We Need GM Crops to Feed the World?," 6 December 2015, https://cban.ca/do-we-need-gm-crops-to-feed-the-world/.

6. Margaret Mellon, *Biotechnology and the Environment* (Washington, D.C.: National Wildlife Federation, 1988).

7. Jane Rissler and Margaret Mellon, *The Ecological Risks of Engineered Crops* (Cambridge, Mass.: MIT Press, 1996).

8. Margaret Mellon and Jane Rissler, *Now or Never: Serious New Plans to Save a Natural Pest Control* (Cambridge, Mass.: Union of Concerned Scientists, 1998).

9. Margaret Mellon and Jane Rissler, *Gone to Seed: Transgenic Contaminants in the Traditional Seed Supply* (Cambridge, Mass.: Union of Concerned Scientists, 2004).

10. Union of Concerned Scientists, *A Growing Concern: Protecting the Food Supply in an Era of Pharmaceutical and Industrial Crops* (Cambridge, Mass.: Union of Concerned Scientists, 2004).

11. Doug Gurian-Sherman, *Failure to Yield: Evaluating the Performance of Genetically Engineered Crops* (Cambridge, Mass.: Union of Concerned Scientists, 2009).

12. Doug Gurian-Sherman and Noel Gurwick, *No Sure Fix: Prospects for Reducing Nitrogen Fertilizer Pollution through Genetic Engineering* (Cambridge, Mass.: Union of Concerned Scientists, 2009).

13. Doug Gurian-Sherman, *High and Dry: Why Genetic Engineering Is Not Solving Agriculture's Drought Problem in a Thirsty World* (Cambridge, Mass.: Union of Concerned Scientists, 2011).

14. Union of Concerned Scientists, *The Healthy Farm: A Vision for U.S. Agriculture* (Cambridge, Mass.: Union of Concerned Scientists, 2013).

15. Gurian-Sherman and Gurwick, *No Sure Fix*.

16. Michael Livingston, Jorge Fernandez-Cornejo, Jesse Unger, Craig Osteen, David Schimmelpfennig, Tim Park, and Dayton Lambert, *The Economics of Glyphosate Resistance Management in Corn and Soybean Production* ERR-184 (Washington, D.C.: U.S. Department of Agriculture, Economic Research Service, April 2015).

17. Charles M. Benbrook, "Impacts of Genetically Engineered Crops on Pesticide Use in the U.S.—the First Sixteen Years," *Environmental Sciences Europe* 24, no. 24 (2012): n.p.

18. David A. Mortensen, J. Franklin Egan, Bruce D. Maxwell, Matthew R. Ryan, and Richard G. Smith, "Navigating a Critical Juncture for Sustainable Weed Management," *BioScience* 62 (January 2012): 75–84.

19. Mortensen et al., "Navigating a Critical Juncture for Sustainable Weed Management."

20. World Health Organization, "Evaluation of Five Organophosphate Insecticides and Herbicides," *IARC Monographs* 112 (20 March 2015): 321–412.

21. Dana Loomis, Kathryn Guyton, Yann Grosse, Fatiha El Ghissasi, Véronique Bouvard, Lamia Benbrahim-Tallaa, Neela Guha, Heidi Mattock, and Kurt Straif, "Carcinogenicity of Lindane, DDT, and 2,4-Dichlorophenoxyacetic Acid," *Lancet Oncology* 16, no. 8 (1 August 2015): 891–92.

22. Jorge Fernandez-Cornejo, Seth Wechsler, Mike Livingston, and Lorraine Mitchell, "Genetically Engineered Crops in the United States," *Economic Research Service Report* 162 (February 2014): 1–54.

23. Benbrook, "Impacts of Genetically Engineered Crops."

24. A. M. Shelton, D. L. Olmstead, E. C. Burkness, W. D. Hutchison, G. Dively, C. Welty, and A. N. Sparks, "Multi-state Trials of *Bt* Sweet Corn Varieties for Control of the Corn Earworm (Lepidoptera: Noctuidae)," *Journal of Economic Entomology* 106, no. 5 (October 2013): 2151–59.

25. Aaron J. Gassmann, Jennifer L. Petzold-Maxwell, Eric H. Clifton, Mike W. Dunbar, Amanda M. Hoffmann, David A. Ingber, and Ryan S. Keweshan, "Field-Evolved Resistance by Western Corn Rootworm to Multiple *Bacillus thuringiensis* Toxins in Transgenic Maize," *Proceedings of the National Academies of Sciences* 111, no. 14 (2014): 5141–46.

26. Environmental Protection Agency, "Proposal to Improve Corn Rootworm Resistance Management," ID-EPA-HQ-OPP-2014-0805-0001 *Federal Register* 80, no. 51 (17 March 2015): 13851.

27. "Corn Soil Insecticide Use Up Dramatically to Combat Widespread Rootworm Challenges," *Agri-View*, 10 January 2013, n.p.

28. Margaret R. Douglas and John F. Tooker, "Large-Scale Deployment of Seed Treatments Has Driven Rapid Increase in Use of Neonicotinoid Insecticides and Preemptive Pest Management in U.S. Field Crops," *Environmental Science and Technology* 49, no. 8 (2015): 5088–97.

29. Dennis Gonsalves, Savarni Tripathi, James B. Carr, and Jon Y. Suzuki, "Papaya Ringspot Virus," in *Plant Health Instructor* (St. Paul: American Phytopathological Society, 2010).

30. Prabhu L. Pingali, "Green Revolution: Impacts, Limits, and the Path Ahead," *Proceedings of the National Academy of Sciences* 109, no. 31 (2012): 12302–308.

31. Gurian-Sherman, *Failure to Yield*.

32. Gurian-Sherman, *High and Dry.*

33. U.S. Food and Drug Administration, "Biotechnology Consultation Note to the File BNF No. 000141. Subject Genetically Engineered (GE) Potato Varieties," 20 March 2015.

34. The International Rice Research Institute (IRRI) reports that as of March 2014 the yields for golden rice were below those for comparable local varieties and that the rice had yet to approved by national regulators or shown to improve vitamin A status under community conditions. See IRRI, https://irri.org/golden-rice, accessed July 2015.

35. Beverly McIntyre, Hans R. Herren, Judi Wakhungu, and Robert T. Watson, eds., *International Assessment of Agricultural Knowledge, Science and Technology for Development (IAASTD): Synthesis Report with Executive Summary: A Synthesis of the Global and Sub-Global IAASTD Reports* (Washington, D.C.: International Assessment of Agricultural Knowledge, Science and Technology for Development), 1–11.

36. Gonsalves et al., "Papaya Ringspot Virus."

37. Bruce E. Tabashnik, Thierry Brévault, and Yves Carrière, "Insect Resistance to *Bt* Crops: Lessons from the First Billion Acres," *Nature Biotechnology* 31, no. 6 (2013): 510–21.

38. John M. Baker, Tyson E. Ochsner, Rodney T. Venterea, and Timothy J. Griffis, "Tillage and Soil Carbon Sequestration—What Do We Really Know?," *Agriculture, Ecosystems and Environment* 118 (2007): 1–5.

39. Vincent Poirier, Denis Angers, Philippe Rochette, Martin Chantigny, Noura Ziadi, Gilles Tremblay, and Josee Fortin, "Interactive Effects of Tillage and Mineral Fertilization on Soil Carbon Profiles," *Soil Science Society of America Journal* 73, no. 1 (January 2009): 255–61.

40. National Research Council, *Impact of Genetically Engineered Crops on Farm Sustainability in the United States* (Washington, D.C.: National Academies Press, 2010).

41. McIntyre et al., *International Assessment of Agricultural Knowledge, Science and Technology for Development.*

42. Gurian-Sherman, *Failure to Yield.*

43. Gurian-Sherman, *High and Dry.*

44. Gurian-Sherman and Gurwick, *No Sure Fix.*

45. Gurian-Sherman, *Failure to Yield.*

46. Gurian-Sherman and Gurwick, *No Sure Fix.*

47. Natasha Gilbert, "Cross-Bred Crops Get Fit Faster," *Nature* 513, no. 7518 (18 September 2014): 292.

48. S. V. Siar, G. A. Beligan, A. J. C. Sajise, V. N. Villegas, and R. A. Drew, "Papaya Ringspot Virus Resistance in Carica Papaya via Introgression from *Vasconcellea quercifolia*," *Euphytica* 181, no. 2 (September 2011): 159–68.

49. Christine Hotz, Cornelia Loechl, Lubowa Abdelrahman, James K. Tumwine, Grace Ndeezi, Agnes Nandutu Masawi, Rhona Baingana, et al., "Introduction of b-Carotene-Rich Orange Sweet Potato in Rural Uganda Results in Increased Vitamin A Intakes among Children and Women and Improved Vitamin A Status among Children," *Journal of Nutrition* 142, no. 10 (1 October 2012): 1871–80.

50. Trudy Bialic, "The Opal Apple: No Browning, Naturally!," *Sound Consumer*, PCC Natural Markets, November 2014, https://www.pccmarkets.com/sound-consumer/2014-11/opal_apple/.

51. Benno Vogel, *Marker-Assisted Selection: A Non-invasive Biotechnology Alternative to Genetic Engineering of Plant Varieties* (Amsterdam: Greenpeace International, 2014).

52. Comment on EPA's Proposal to Improve Corn Rootworm Management, Docket ID # EPA-HQ-OPP-2014 (25 March 2015).

53. Heriberto Cerutti and J. Armando Casas-Mollano, "On the Origin and Functions of RNA-Mediated Silencing: From Protists to Man," *Current Genetics* 50, no. 2 (August 2006): 81–99.

54. Jonathan G. Lundgren and Jian J. Duan, "RNAi-Based Insecticidal Crops: Potential Effects on Nontarget Species," *BioScience* 63, no. 8 (1 August 2013): 657–65.

55. T. Gaj, C. A. Gerschback, and C. F. Barbas, "ZFN, TALEN, and CRISPR/Cas-Based Methods for Genome Engineering," *Trends in Biotechnology* 31, no. 7 (2013): 397–405.

56. Yuriy Federov, Emily M. Anderson, Amanda Birmingham, Angela Reynolds, Jon Karpilow, Kathryn Robinson, Devin Leake, William S. Marshall, and Anastasia Khvorova, "Off-Target Effect by siRNA Can Induce Toxic Phenotype," *RNA* 12 (2006): 1188–96.

57. Heidi Ledford, "CRISPR, the Disruptor," *Nature* 522 (2015): 20–24.

2

Born in the U.S.A.

The Americanness of Industrial Agriculture

PETER A. COCLANIS

These are fast times for foodies, salad days for farmers markets, boutique producers, locavores, vegans, and the like. The words "local," "organic," and "slow" have assumed the status of mantras among many pious eaters, ostensibly facilitating spiritual transformations, even as "local," "organic," and "slow" in the real world generally translate into inflated food prices and plummeting agricultural productivity. There is a Food Network on TV; a lavish and lush journal devoted to food, *Gastronomica*; and wildly popular websites such as yelp.com, whose reviews can make or break restaurants anywhere in the country. The study of obesity has become phat, and an entire new book genre has emerged: commodity studies. Virtually every grain, spice, drink, and tuber now has its historian. I can't get too snide here, of course, for I myself am working on rice! And in Chapel Hill, where I live, when the abbreviation "CSA" is mentioned, more people—dollars to doughnuts (Krispy Kreme or otherwise)—think of Community-Supported Agriculture than of the Confederate States of America. Who would have thunk it?

Not long ago, things were far, far different. Indeed, in the late 1980s agriculture seemed like yesterday's news and the study of agricultural history atavistic. Around that time I began compiling a list of stories, anecdotes, and assorted (sordid?) facts suggestive of the diminishing hold of agriculture on the cultural imagination in this advanced postindustrial country. In 1988, for example, the venerable farm youth organization Future Farmers of America, founded in 1924, officially changed its name to FFA, in part to appeal to a broader "food system" constituency, in part, I suspect, because farming qua farming seemed off-putting and somewhat déclassé to modern teens. Interestingly enough, just a few years later—in 1991—Kentucky Fried Chicken officially changed its corporate name to KFC for much the same reasons: the company didn't want to be associated too closely with the adjective "fried," which at the time was becoming similarly déclassé, at least among some small but influential segments of the population. One intriguing, maybe even hopeful countertrend: in 2007, the company began using the old name, Kentucky Fried Chicken, as part of a corporate rebranding process in the United States. We'll see where it leads: stay tuned.

On the whole, though, the evidence I kept coming across suggested that farming and farmers were long ago and far away. In 1998 a graduate-student columnist wrote in the *Daily Tar Heel*, UNC–Chapel Hill's student newspaper, that college was the time for young people to "sew" their wild oats, spelling the infinitive *s-e-w*. The troubling fact that the grad student was in history should be noted as well. In that same year, the actress Mary Frann—best known for her role as Bob Newhart's wife on the TV show *Newhart*—died. In her *New York Times* obituary, a friend was quoted as saying that in the last few years before her death, Frann had had "a hard road to hoe." Asphalt or concrete would prove quite difficult to hoe, I suppose.

Later in that same year, Auburn University changed the name of its Department of Agricultural Engineering to the Department of Biosystems Engineering. The *Wall Street Journal* ran an article entitled "Auburn Seeks to Revamp Aggie Image" specifying the reasons for the name change. According to the piece, the old name wasn't very appealing to prospective students, who increasingly viewed agriculture in negative terms as "an unsophisticated, low-technology field." When an institution such as Auburn—Alabama's land-grant school since 1874—is embarrassed by its connection with farming, you know agriculture is in trouble.[1]

Before quitting the 1990s, one last gem. I myself recall, ruefully, a breakfast gathering sponsored by the Agricultural History Society at a meeting in Chicago of the Organization of American Historians. The crowd at the meeting was modest. All ten or so of us in attendance were middle-aged or aged white males—although I promise I was not wearing a string tie like two or three others at the breakfast. Anyway, the meeting was going full bore (if you know what I mean) when two very chic, well-dressed young women slipped in and sat down at adjacent empty seats. They sat for about ten or fifteen minutes, listening politely to us old white guys, when one of them asked, again very politely, "Is this the Oral History Association breakfast?" Alas, we said no, and helpfully pointed them to a room across the hall and they were off, leaving us to gum the rest of our oatmeal and transact other difficult business in the manner in which we had long—too long—been accustomed.

As I suggested at the outset, however, over the last fifteen years or so things have changed—at least on the surface. Studying food has become hot. Work on agriculture and agricultural history has picked up, even though much of it relates not to the agricultural mainstream, but to fringe groups and movements, outliers and anomalies, alternative practices and traditions, and what is rather more environmental history than agricultural history per se. And although interest in agriculture has grown of late, there is still evidence out there that not everyone has gotten the message. In 2002 I was second reader on a senior honors thesis in history, the writer of which stated at one point that "in a capitalist society . . . one man weeps, the other sows."[2] At least he spelled sow *s-o-w*. A few years back one of my senior colleagues in history at UNC–Chapel Hill told me that he leaves farmers completely out of his U.S. History since 1865 survey because they interfere with his course's narrative thrust; and historian Louis Ferleger has recently demonstrated that coverage of agriculture (particularly southern agriculture) is steadily decreasing in U.S. history texts. And in 2010, the American Farm Bureau Federation (a.k.a. the Farm Bureau), then in its ninety-first year of existence, sold off its web domain name, FB.com. Since the buyer was Facebook, and the price $8.5 million, we can perhaps cut the FB—the largest general farm organization in the United States—some slack.[3] But no breaks for the historians!

So what is the upshot of these introductory remarks? Just a throat-clearing exercise or is there deeper intent? Although the author is never the sole arbiter of these things, I am opting for the latter interpretation. I started this piece by juxtaposing the rising interest in food, food studies, and certain epiphenomenal features of agriculture with some evidence regarding the lack of knowledge about or at times even interest in the basics of American agriculture. Why? Because disarticulation of this sort has deleterious consequences, perhaps the most serious of which for our purposes relate to interpretative distortions of various kinds about American agriculture's history, present condition, and future prospects. Not that the surging interest in food is a bad thing in and of itself, but without deeper grounding, the outgrowth of such interest, whether intentionally or unintentionally, can be misleading and result in serious misinterpretation. In this regard the landscapes of Henri Rousseau, with their distortions in scale and perspective, are somewhat analogous: they are enriching, offering insights, but not closely related to material realities, to the facts on and in the ground, as it were.

With these points in mind, what I hope to do in this chapter is to make the case that despite the relative lack of interest today in normative dimensions of American agriculture or in core themes in American agricultural history, such topics are still important for many, many reasons. The most notable of these reasons is that America's development into a postindustrial superpower, the world's hyperpower to use a formulation the French like to throw around, has been inextricably connected to our country's formidable record of success in the agricultural sector over the centuries. My argument comprises four parts. First, I'll try to generalize a bit about the historical character of American agriculture and the principal factors responsible for the robust performance of the American agricultural sector over the past 350 or so years. The contours of American agriculture, as it were. Second, I'll present some historical data illustrating the long-term decline in the relative importance of the American agricultural sector. Third, I'll make the case that despite its declining relative importance the agricultural sector has proved central to American urbanization and industrialization, indeed, to the process of economic development more generally in the United States. Fourth, I'll try to contextualize the (growing) place of alternative farming in the American agricultural system. In so doing I'll spell out some of the impediments to its expansion and establish its proximate bounds and limits.

The readers of this volume all are interested in food and agriculture and likely have some degree of commitment to sustainable development. Nonetheless, it is important to recall now and then that even in our voluntaristic times there are a range of structural factors that circumscribe our actions, however well-intended, if at times unrealistic and, I believe, self-righteous they may be. In other words, then, with respect to food, Marx may have had it about right when he wrote in *The Eighteenth Brumaire of Louis Bonaparte* that "men make their own history, but they do not make it just as they please."[4]

With these considerations in mind, let us turn to the subject at hand: the contours of American agriculture, so to speak. First, to the explanatory factors, or, more properly, sets of factors most responsible for our agricultural success over time. And here I'll be adopting the rhetorical style of economics rather than history, that is, privileging analysis over narration, patterns over variations, and explanatory parsimony over thick description. When we employ this m.o., four categories of variables come immediately to mind in explaining our agricultural success over time: (1) resources; (2) markets; (3) sociocultural values; and (4) institutions. Generally speaking, all four categories of variables were extremely propitious for agricultural success throughout most of our history; taken together, the patterning of these variables goes a long way toward explaining America's position of power and preeminence in the world today.

America's resource "endowment" has been exceptionally favorable for agricultural development throughout our history. With a huge inventory of fertile, well-watered, and varied land, a person-to-land ratio very conducive to productivity and prosperity, good access to financial capital most of the time, and a heterogeneous population mix—European, African, Native American, Asian—possessed of different and often complementary forms of proprietary knowledge about farming, America's agricultural sector has generally been extremely well-positioned for farm-building, technological advance, and development.[5]

For much of our history, moreover, the conjuncture of supply and demand possibilities for most American agriculturalists has been conducive to growth. With markets widening and deepening in the West over the past three centuries as a result of increases in population and levels of urbanization and wealth, improvements in transportation, storage, distribution, and communication, and the greater commercialization of human values, demand for American foodstuffs has generally been strong, not only domestically, but often internationally as well. It should be noted, moreover, that in recent decades much of the growth in international demand for American farm products has come from Asia. Even as Engel's law began to kick in and American families spent a smaller percentage of their income on food when their incomes increased, American farmers were able to adjust by shifting their output mixes to varying degrees toward agricultural products of greater income elasticity (higher-quality meats, dairy products, and specialty fruits and vegetables).[6] The result of the conjuncture of strong demand and American farmers' relatively elastic supply response—we shall outline some of the reasons for the elasticity of supply momentarily—has generally been a high-output, high-price equilibrium in American agriculture over sustained periods of time in U.S. history.[7]

The elasticity of agricultural supply in the United States has historically been related closely to our last two categories of explanatory variables: sociocultural values and American institutions, both of which have helped to condition responsiveness to market signals and signs. From the time of initial colonization and settlement, American farmers, by and large, have been of an entrepreneurial bent and have responded pretty vigorously and in what conventional economists would refer to as rational ways to price signals. There was seldom a backward-bending labor supply curve in American agriculture, and most American farmers have pursued utility-maximizing (if not profit-maximizing) strategies with method and vigor whenever possible. In so doing, they have been helped by the institutional framework in which they found themselves embedded: a framework, that is to say, informed by strong legal support for contracts, private property, and individual liberties and freedoms, and concerted governmental support for prodevelopmental policies, broadly conceived (including agricultural research and infrastructure). Indeed, the entire institutional apparatus in America—constitutional arrangements, property laws, court decisions, governmental fiscal and monetary policies, efficiency-enhancing levels of regulation, and so on—has proven hospitable to bold developmental initiatives by farmer-entrepreneurs (nascent agribusiness people) across the United States.

In America, then, we find enterprising farmers with adequate farm finance situated on fertile, but relatively inexpensive land, operating for the most part in a favorable market environment with strong institutional support from the legal system and political/governmental order. Such farmers were aided over time by increasingly taut supply chains marked by ever-improving logistical systems and storage/processing facilities, and by increasingly sophisticated financial/insurance mechanisms, the last of which served important risk-reduction and income-smoothing functions.[8] In such an economic and institutional climate, it is not surprising that American farmers have been extraordinarily productive over time, regularly bringing in impressive "harvests" of marketable crops and livestock via increasingly efficient means. In so doing, they have enriched themselves, gratified consumers, and greatly facilitated the economic development of the United States. Even when taking into consideration the negative externalities associated with agriculture—ground and water pollution, aquifer depletion, and so forth—one cannot gainsay the fact that over the centuries the performance of the farm sector on balance has proved a terrific boon to the United States.

The startling changes in the role and position of America's agricultural sector over time are captured well in some data, arrayed in tables 1 and 2, on agriculture's share of the American labor force and agriculture's share of U.S. domestic product. Although these data are rough, they do establish some broad levels of magnitude.

As these data demonstrate, agriculture's share of both the labor force and gross domestic product has declined dramatically over the past two centuries. Of course, in absolute terms—whether measured in dollars or in sheer quantity—the market value of agriculture, food, and related

TABLE 1. AGRICULTURE'S SHARE OF U.S. LABOR FORCE

Year	Percent
Ca. 1750	75–85
1800	74.40–83.30
1850	55
1900	40.02
1950	12.04
2016	1.37*

*Workers involved in crop production, animal production, and aquaculture
Sources: Edwin J. Perkins, *The Economy of Colonial America* (New York: Columbia University Press, 1980), 41; Edwin J. Perkins, *The Economy of Colonial America*, 2nd ed. (New York: Columbia University Press, 1988), 57; Stanley Lebergott, *The Americans: An Economic Record* (New York: W. W. Norton, 1984), 66, Table 7.3; Thomas J. Weiss, "U.S. Labor Force Estimates and Economic Growth, 1800–1860," in *American Economic Growth and Standards of Living before the Civil War*, ed. Robert E. Gallman and John Joseph Wallis (Chicago: University of Chicago Press, 1992), 19–78, esp. 22, Table 1.1; *Historical Statistics of the United States, Earliest Times to the Present, Millennial Edition*, ed. Richard Sutch and Susan B. Carter, 5 vols. (New York: Cambridge University Press, 2006), 2:101, Table Ba652–669; U.S. Department of Labor, Bureau of Labor Statistics, Labor Force Statistics from the Current Population Survey, Household Data, Annual Averages,18b, Employed Persons by Industry and Age, last modified February 8, 2017, https://www.bls.gov/cps/cpsaat18b.htm, accessed May 18, 2017.

industries is greater today than ever before, reaching an all-time high of $992 billion—5.5% of GDP—in 2015.[9] What may be less clear, at least on the surface, however, is the manner in which agriculture has supported, underpinned, and reinforced urbanization, industrialization, and American development more generally even as farmers and farming have receded both from the fields themselves and from the American historical imagination.

There are, at the very least, four important ways in which agriculture has played a crucial role in shaping, if not conditioning or even determining the pace and pattern of American economic expansion over the last few centuries.[10] First, because of our favorable factor endowment and the steady stream of productivity gains in agriculture over time, food and fiber prices in America have generally been very low in comparative terms. As a result, Americans have traditionally spent a very modest proportion of disposable personal income on food—in 2014 the figure was less than 10%, with about 5.5% spent on food consumed at home, and another 4.3% on food consumed away from home. In percentage terms, Americans spend less of their incomes on food than almost any nation in the world. Most people in developed countries spend twice as much in percentage terms, and in less-developed countries (LDCs) four to eight times as much.[11] For the United States, low food

TABLE 2. AGRICULTURE'S SHARE OF U.S. GROSS DOMESTIC PRODUCT

Year	Percent
1800	46
1840	40
1870	33
1900	18
1930	8
1945	7
1970	2
2006	0.9
2012	1
2015	1

Sources: Marvin Towne and Wayne Rasmussen, "Farm Gross Output and Gross Investment in the Nineteenth Century," in *Trends in the American Economy in the Nineteenth Century*, ed. William Parker, Conference on Research in Income and Wealth, National Bureau of Economic Research, vol. 24, *Studies in Income and Wealth* (Princeton, N.J.: Princeton University Press, 1960), 255–312, esp. 265, Table 1, Farm Gross Product, Decade Years, 1800–1900; *Historical Statistics of the United States, Earliest Times to the Present,* Millennial Edition, ed. Richard Sutch and Susan B. Carter, 5 vols. (New York: Cambridge University Press, 2006), 3: 23–28, Table ca9–19, Gross Domestic Product: 1790–2002, esp. 23; Robert E. Gallman, "Economic Growth and Structural Change in the Long Nineteenth Century," in *The Cambridge Economic History of the United States*, ed. Stanley L. Engerman and Robert E. Gallman, 3 vols. (New York: Cambridge University Press, 1996–2000), 2 (2000): 1–55, esp. 50, Table I.14, The Sectoral Distribution of GNP 1840–1900; Carolyn Dimitri, Anne Efflund, and Neilson Conklin, *The 20th Century Transformation of U.S. Agriculture and Farm Policy*, U.S. Department of Agriculture, Economic Research Service, Economic Information Bulletin, Number 3, June 2005 (Washington, D.C.: 2005), 2; U.S. Department of Agriculture, Economic Research Service, Chart Gallery, *What is Agriculture's Share of the Overall U.S. Economy?*, November 25, 2014, http://www.ers.usda.gov/data-products/chart-gallery/detail.aspx?chartId=40037&embed=True, accessed January 20, 2015; "Agricultural Output Climbed in 2013, Recovering from Drought," *New York Times*, June 20, 2014, http://www.nytimes.com/2014/06/21/business/economy/after-a-drought-agriculture-climbs.html?_r=0, accessed June 30, 2015; U.S. Department of Agriculture, Economic Research Service, *Ag and Food Sectors and the Economy* [2015], https://www.ers.usda.gov/data-products/ag-and-food-statistics-charting-the-essentials/ag-and-food-sectors-and-the-economy.aspx, last updated May 5, 2017, accessed May 18, 2017. The author would like to thank Paul W. Rhode of the Department of Economics at the University of Michigan for help in calculating the estimate for 1800.

costs mean that a large proportion of American income is "freed" for other uses, whether for consumption of manufactured goods and services, leisure activities, or investment of one type or another. To be sure, some would argue that this percentage is too low, that negative externalities need to be factored in via environmental accounting, but I assure you, almost every other country in the world would love to start with our alleged problem.

Second, because many of the productivity gains in America have been labor-saving (whether through mechanization or productivity-enhancing biological inputs such as improved seeds, herbicides, and pesticides), many workers were rendered redundant relatively early on in the agricultural sector. Although this often posed assorted short-term traumas and difficulties for the affected workers themselves, technological change of this sort helped create the labor pool necessary for America's massive industrialization process between 1850 and about 1960.

Third, the agricultural sector has proven to be a strong source of demand for American industrial goods over time. Indeed, many urban factories across the United States have kept busy producing machinery, implements, tools, biological inputs, and transport, storage, and distribution facilities deployed in/for the farm sector.

Finally, another huge part of the U.S. manufacturing sector has traditionally been involved in the processing of food and fiber produced on American farms. Many of our biggest industries, historically and even today—meatpacking, other food products, textiles, paper and pulp, liquor and alcohol, leather, and so on—are all essentially processing industries for raw materials produced in America's farm sector.

If the American experience reveals anything at all about the general process of economic development, it has, in fact, been the importance of the linkage, the organic interaction, as it were, between the urban and rural sectors, between factories and fields. In no place was this truer, historically, than in the so-called manufacturing belt of the United States: broadly speaking, the northeastern quadrant of the United States (the area east of the Mississippi River and north of the Ohio River). In this region, commercial, agriculture developed simultaneously with cities and manufacturing. Over time, as agriculture developed (sending agricultural surpluses to the region's villages, towns, and cities), the urban centers and factories in the region began to supersede outside suppliers of manufactures to meet the needs of the region's increasingly prosperous farmers as well as other denizens in the region. Moreover, as the region's farmers and manufacturers improved in productivity and grew in scale over time, they themselves began to export their products successfully to other parts of the United States and abroad. In so doing, this process brought sustained prosperity to the region as a whole, creating what has sometimes been referred to as our agro-industrial complex. The agro-industrial complex, once created, spread to the Great Plains, before jumping to California, a state that in agricultural terms drew inspiration from—and in many ways came to resemble—Iowa and other midwestern states for a considerable period of time.

For a century and a half this growing complex was the envy of the world and in some ways still is. If the experience of the manufacturing belt was not exactly a textbook case of what some development experts today call Agricultural Development-Led Industrialization (ADLI), it was at the very least an example of a "virtuous economic circle" established through the simultaneous and balanced interactions between market agriculture and regional urban and industrial constellations.

The case of the U.S. manufacturing belt, and, to some extent, the case of America more generally, should give pause to historians, economists, and planners devising developmental strategies even today. The position of agriculture may no longer be as visible here as it once was, but without a prosperous rural sector—which in my view itself depends on a stable institutional setting with enforceable property rights—sustained and sustainable industrialization, let alone development anywhere will be less a reality than a hope and a dream.[12] It is a hopeful sign, then, that over the past few years, two of the world's most powerful forces—the World Bank (whose 2008 development report focused on agriculture) and the Gates Foundation—have in the first case rediscovered agriculture and in the second case started paying systematic attention to it for the first time.[13]

Now let us situate alternative agriculture and what some call the L-O-S movement—local, organic, and slow—in this historical context. In order to do so, let me start, perhaps a bit counterintuitively, by reiterating and reemphasizing the close historical relationship in America between agriculture and industry. Indeed, if we take our cue from historians of technology and define industrialization broadly—and not equate it with machines or even with manufacturing—what we need to focus on is a particular way of thinking about and acting in the material world. Conceived in such terms, industrialization can profitably be defined as the achievement and institutionalization of historically high rates of productivity through the systematic employment of scientific knowledge to transform the material environment into a flow of economically useful goods and services.[14] Over time, the employment of such knowledge in the material world has led to drastic increases in economic productivity (the relationship of inputs to output), which is the key to growth and development. Note also that when industrialization is conceived of in this way, one can talk without contradiction of industrial agriculture emerging simultaneously with—or even as a precursor of and catalyst for—the changes in mining and manufacturing technology typically associated with the so-called Industrial Revolution.[15]

In America we can see the roots of industrial agriculture centuries ago. Indeed, the industrialization of our economy as a whole began early on, and the various sectors over time have shared many of the same characteristics. For example, the first "big business" in America, as Alfred D. Chandler Jr. pointed out long ago, was the southern plantation, and many of the organizational and managerial strategies and structures developed on large plantations—specialization, division of labor, scale economies, middle management positions, consolidated accounting—were adopted and refined in the manufacturing sector later in the nineteenth century.[16] Abstracting a bit, it is undeniable that America's most original historical contributions to production technology over time have involved large units, and the efficient, capital-intensive mass production of standardized products of low/average quality at cheap prices. This is so whether we are talking about the production of milling machines in New England in the early nineteenth century or cotton production in the antebellum South, or about the "disassembly" lines in the Cincinnati and Chicago meatpacking industries in the second half of the nineteenth century, wheat production on the bonanza farms of North Dakota in the 1880s and 1890s, or Model Ts at Henry Ford's plant in Highland Park, Michigan, in 1914.[17] Or about the labor-deskilling innovations in meatpacking introduced in the Midwest in the 1980s by IBP (now Tyson Fresh Meats), the bête noire of labor unions, or about the "assembly" of food at Taco Bell, Applebee's, or Mickey D's.[18] What we traditionally do well is bulk not batch, quick not slow, cheap not luxury, big box not boutique. Our "industrial agricultural" system today involves much the same thing, to the chagrin of Eric Schlosser, Morgan Spurlock, Michael Pollan, Ted Genoways, and Barry Estabrook—and to the pocketbook advantage, if not delight, of many hard-pressed consumers at home and abroad.

Now why do I pay so much attention to these matters in a section of a chapter ostensibly devoted to alternative agricultures and L-O-S? Simply put, in order to get readers to think realistically about food politics—politics in a Bismarckian sense as "the art of the possible." To underscore this point, let's do the numbers, as *Marketplace*'s Kai Ryssdal might say. While boutique farms and farmettes are all the rage with food romantics, the mean size of an American farm in 2012 was 434 acres (168 in my home state of North Carolina).[19] Size and scale matter in American agriculture. Over 64% of cropland harvested is produced on farms of over one thousand acres, and almost 80% on farms of five hundred acres or more.[20] The vast majority of American farms make little difference in the marketplace. The smallest 75% of farmers accounted for only 3% of the value of agricultural products sold in 2012, with 97% of farm sales in that year coming from the largest 25%. And that's not the half of it. The largest 4% accounted for 66% of the total value of agricultural output in 2012, with the top 0.5% alone accounting for 32%.[21]

The small farmers at my local farmers market notwithstanding, boutique farms are in the aggregate pretty much irrelevant to American agriculture as a whole. Most small farmers today derive very little income from agriculture, in any case: almost all income of so-called "farm" households actually comes from nonfarm employment of one kind or another. Indeed, even considering all farmers in America—that is, even including large ones—on average only 11% of household income for farm families in 2010 was estimated by the USDA to come from agriculture.[22] Without the largest farms, in other words, America doesn't really have a significant commercial agricultural sector, certainly not one capable of both feeding the U.S. population affordably and exporting over $135 billion worth of agricultural products in calendar year 2016, near our greatest total ever.[23] So while I'm really not trying here to channel Earl Butz, who as U.S. secretary of agriculture in the 1970s famously challenged American farmers to "get big or get out," I am trying to inject a much-needed dose of realism into debates over farming in this country.

While many of us in North Carolina are enamored of our farmers markets, our custom producers, and our CSAs, what the Old North State is important for, agriculturally speaking, is its pioneering factory hog farms ("hog integrators") and being among the nation's leaders not only in hog production (#2), but also in the production of tobacco (#1), sweet potatoes (#2), poultry and eggs (#2), turkeys (#3), Christmas trees, and cucumbers; it's a leading producer of strawberries, bell peppers, upland cotton, and broiler chickens as well.[24] North Carolina is also important for being the site of the Open Grounds Farm, which I doubt that many readers of this piece have heard of. This low-profile, Italian-owned farm in Carteret County near the coast produces huge quantities of corn, soybeans, and winter wheat, and at roughly fifty-seven thousand acres in size is the largest farm east of the Mississippi River.[25] Again, who would have thunk it?

At this point, some readers might be asking what about organic agriculture? To be sure, the sector has grown rapidly since the 1980s, albeit from an infinitesimal base, and organic farmers on average are younger than farmers practicing conventional agriculture. But in a macro sense organic agriculture even today comprises a tiny share of American agriculture and is of only marginal importance to most Americans. To be more specific, according to the USDA's National Agricultural Statistical Service, organic food accounted for about 5.5% of the total value of agricultural sales in the United States in 2014.[26]

And from my perspective, it's not a bad thing that the organic sector is tiny, because if it grew much larger it would not only raise food costs and reduce the overall efficiency of the agriculture sector but would also lead to potentially catastrophic environmental problems. As the distinguished agricultural expert Robert Paarlberg of Wellesley College pointed out in 2010 in an influential article in the journal *Foreign Policy*, a complete switch-over to organic agriculture would be nothing short of disastrous.[27] To replace the synthetic nitrogenous fertilizer currently used on American farms with "natural" organic fertilizer, we'd have to increase the U.S. cattle stock fivefold. If these animals were raised on organic forage, we'd have to convert much of the land in the lower forty-eight states to pasture. Because the yield of organic field crops is generally much lower on average than nonorganics—studies published in journals such as *Nature* and *Science* and data compiled by the U.S. Department of Agriculture put the shortfall at between 5% and 38% for various crops under different tillage conditions—much more land would have to be converted to crops.[28]

Paarlberg estimates that if Europe tried to feed itself organically, it would need an additional 28 million hectares (one hectare is equal to 2.47 acres) of cropland—69 million acres—equal, in his words, "to all the remaining forest cover in France, Germany, Britain, and Denmark combined." And why? Probably not for nutritional purposes. Although one can certainly find studies proclaiming the nutritional superiority of organic food, the most comprehensive study ever done on this subject was published in 2010 in the *American Journal of Clinical Nutrition*. In this study the research team conducted a "systematic review" of studies done over the past fifty years on the health benefits of organics, finding no nutritional advantage of organics over conventionally grown food. Other studies, including studies done by the Mayo Clinic and by Stanford University researchers, largely concur.[29] Moreover, according to the FDA, even the highest dietary exposures to pesticide residues found on foods grown conventionally in the United States are so small—less than one one-thousandth of toxicity levels—that the purported safety gains from buying organic are trivial. To be sure, exposure to pesticides remains a serious problem in the LDCs, where pesticide use is not well regulated, but, according to Paarlberg, even there the problem is mainly for the unprotected sprayers rather than for consumers.[30]

Ironically, if organics ever become a major commercial presence in the United States, it will likely be due to "big box" stores, including Walmart, the bugbear of food snobs everywhere. In a highly publicized move a few years back, the Bentonville behemoth made a commitment to ramp up its organic offerings and to sell them at reasonable prices. Mon dieu! It is now one of the largest sellers of organics in the United States—either first or second behind Costco—but we'll have to wait and see what the future holds.[31]

And what about the production of meat in small-scale, nonindustrial settings? According to a piece a few years back by James E. McWilliams—a vegan critic of both meat-eating and industrial agriculture—this is a nonstarter too, for the negative environmental impact of small-scale production is even greater (particularly regarding global warming), small-scale production is not necessarily "more natural," and the economics of small-scale production are untenable.[32]

In the face of all this, people like Michael Pollan still want us to go local and organic. In 2010 he upset even some of his fans when he put some specific numbers behind his "pay more, eat less" mantra, and stated that we should be willing to pay $8 for a dozen eggs and $3.90 for a pound of peaches![33] Numbers like that make me wonder just what type of organic output Pollan is ingesting!

In 2006 Dave Eggers published an interesting novel titled *What Is the What*, which leads me to ask of Pollan, who is the "we"? In a country wherein the median income has basically been stagnant for thirty years—where, in 2015, 13.5% of the population (43.1 million people) was living beneath the poverty line—just who does he mean?[34] In this regard, it is worth noting as well that over the past twenty-five years, the bottom quintile (20%) of households has spent between 29% and 46% of household income on food.[35] In unsettled times like this, many, if not most, Americans are thankful that this country's food—produced industrially—is so cheap, at least at Walmart, Costco, Lidl, and Aldi, if not at Whole Foods. Most of the world's population would relish the opportunity of dealing with the problem of food that is too cheap.

Now I'm being intentionally provocative here. There are many problems with what some call scientism, and with the type of top-down technocratic approaches—what James Scott calls "high modernism"—that have shaped our industrial agricultural system over time.[36] I don't particularly like the way that chickens are raised either, for example. And in this regard I am certainly joined by many other Americans. Indeed, in this volume, Steve Striffler has carefully laid out both the calls for reform and the outright opposition that industrial agriculture have often occasioned. But the answer is not to apotheosize preindustrial agriculture and to sanctify local-organic-slow regimes, as Ken Albala seems to do in his contribution to this volume, but to improve and render more humane our industrial agricultural system (which is currently being done, by the way). You want local, organic, and slow? The closest thing to that regime in the world, as Paarlberg points out, is that prevalent in Africa, the least productive agricultural regime in the world, a place whose largely L-O-S regime hasn't been able to feed its population of a billion people in decades.[37]

No, the answer in my view is not boutique L-O-S agriculture—however beneficial to small, well-heeled populations and to gentleman farmers–cum–Jefferson lecturers such as Wendell Berry—but rather a better, more egalitarian and accommodating industrial regime, one still predicated on high technology and the systematic employment in the agricultural realm of the best scientific knowledge.[38] Over the past fifty years developments in this regime have brought us not only the Green Revolution, which allowed the world to cope with massive increases in population in the 1960s, 1970s, and 1980s, but also, more recently, impressive biological innovations (GMOs, most notably), and so-called precision farming—with "no-till" planting, drip irrigation, advanced agricultural analytics, GPS/sensor systems for chemical dispensing, and so on—that together have significantly reduced farming's environmental footprint. The OECD's 2008 review of the "environmental performance of agriculture" in the world's thirty most advanced countries—the countries characterized most by capital-intensive, high-tech agriculture—demonstrates that production had grown substantially between 1990 and 2004, while negative environmental impacts had fallen in every category and biodiversity had increased. And subsequent updates have shown that progress has continued since that time.[39] And many other studies have demonstrated that the food systems of the OECD countries are the safest in the history of the world, whatever people such as Mark Bittman would have us believe. As I pointed out in a piece in the *Wall Street Journal* in 2011, in this country one's chances of dying from ingesting bad food or water at an individual "eating event" is about 1 in 125 million: pretty good odds, it seems to me.[40]

The biggest food, agriculture, and sustainable development problem we will face over the next two generations is not that posed by bad food, cheap food, excessive food-miles, or the horrors of HFCS (high-fructose corn syrup)—oops, corn sugar—although each of these poses formidable challenges in its own way. The biggest problem, as the FAO, the UNDP, and other international agencies know well, is somehow finding a way to feed the roughly 9.7 billion people who will be living on earth in 2050—a third more than are living on earth today—on less land, worse land, with less water, fewer herbicides and pesticides, and almost certainly in a higher-temperature climatic regime. The fact that this population, on balance, will be wealthier and likely demand more foods of higher income elasticities (meats, dairy products, and the like) will make the above challenge even greater.[41] Clearly, many adjustments will have to be made—changes in diet, and reductions in food waste come immediately to mind—if we are to be food secure in 2050. But in my view food security will come about—if it does come about—mainly via the continued development of high-technology, scientific, industrial agriculture, which will certainly include GMOs, gene editing through techniques such as CRISPR-Cas9, synthetic biology, and microbiomes, among other things.[42] The fact that there is at least some possibility that in the future industrial agriculture will literally be sited in factories—European and American biotech companies are currently experimenting with lab-grown meat and milk—adds a mordant, even sardonic twist to things. Although the last scenario isn't all that appealing on the surface—even to me—the African adage "Human rights begin with breakfast" trumps aesthetics, especially if nutritious lab-produced food uses fewer resources and is priced cheaply.[43]

One must, of course, always be wary of the exaggerated claims often associated with new technology, scientific "breakthroughs," and so on—as Margaret Mellon makes clear in her insightful essay in this volume—but, as an economic historian, I ask where precisely would we be today without earlier episodes of "scientific progress" in agriculture? Without irrigation? Without crop rotation and clover? Without the Mendelian revolution? Without hybrids? Without Haber-Bosch? And, in my view, without GMOs, gene editing, and the like? Where? Simply put, L-O-S, or, more to the point, LOST! It is possible, of course, that in the future we might be able to bring together in more significant ways GMOs and organics—as Pamela C. Ronald and Raoul W. Adamchak have provocatively proposed—but, if it happens, it will for the most part be scientific progress that will bring such a melding about.[44]

Notwithstanding the above considerations, in late October 2010 a major private research university in Durham, North Carolina, celebrated "Food Week," but found no place on the program for issues relating to the need to increase food production or for any consideration of mainstream agriculture per se. The program found time for a Duke Iron Chef 2010 competition, using pumpkins and "sustainable ingredients," but no place for discussions of the pros and cons of Monsanto's Roundup Ready seeds (or of glyphosates, for that matter) or of so-called golden rice, enriched with beta-carotene (vitamin A).

Hope springs eternal and is sometimes rewarded. In February 2012 another food conference was held in the Triangle area—"Shared Tables"—that included a bit more balance regarding speakers and perspectives. To be sure, its promotional materials also included a precious recipe for "rosemary garlic goat cheese scones," and after the festivities the conference website included the following self-righteous and self-indulgent proclamation: "The conference also had a practical impact on the environment as we composted a total of 16.5 bags of waste, while only sending 4 bags of garbage to the landfill. Additionally, many attendees utilized public transportation, and carbon offsets were purchased for those who travelled to attend." Let's give it up for the Triangle's bourgeoisie![45]

And then there are events like the conference out of which this volume grew—dealing with serious issues relating to food in an open way, no decks stacked, no party line, no PC. Which is as it should be for food systems are so complex that multi- and cross-disciplinary approaches, a variety of perspectives, and, above all, balance and realism—a sense of the possible—are needed if real progress is to be made regarding the challenges ahead. Thus, essays in this volume by scholars with views so different as Striffler and Mellon, on the one hand, and yours truly, on the other. In essence, then, what I'm getting at here is that while we must acknowledge the important contributions made by people such as Alice Waters, Marion Nestle, and, yes, Michael Pollan, we must acknowledge too that we live in a world of cheese fries and flavor-blasted Doritos, of Roundup, of GMOs, and, if I had my way, of DDT.[46] And a world, I can't help but point out, wherein a recent, well-crafted survey found that about 80% of Americans surveyed support "mandatory labels on food containing DNA."[47] Go for it.

Another case in point: Clearly, it is important to note that a small proportion of American adults are vegetarians or vegans: 3%, according to a 2008 report in *Vegetarian Times*, and 7%, according to a 2012 survey conducted by Gallup. Fair enough, but it is also important to remember that the other 93–97% eat meat.[48] While there is a place in our world for edible schoolyards, for locavores, for food purists, and, yes, for food snobs, there should be places, too, for the pretty good and even for the not-so-good, especially when reasonably priced. In other words, places for the expanding French (semi-fast-food) café/bakery chain Paul, for In-N-Out Burger, and even for the much-criticized Hostess Twinkie, which, after a brief absence from the market, returned with a splash in 2013.[49]

The food world is complicated, in other words. Who in 2017 hasn't heard the story of Jared Fogel, the (in)famous and now-imprisoned Subway Guy? How else can one explain the fact that in 2010 a college professor—indeed, a professor of human nutrition at Kansas State University—spent two months on a "convenience store" diet composed for the most part of Oreos, Doritos, and Twinkies, and in so doing lost twenty-seven pounds, improved his health, and got his BMI down to the normal range? Or the Iowa science teacher who in 2013 lost thirty-seven pounds and saw his cholesterol level drop significantly after eating nothing but McDonald's food for three months? Or consider the robust health and physical stature of former UNC football star Bruce Carter, now on the New York Jets, who, according to *ESPN Magazine* in 2010 was arguably the most gifted athlete in college football, a player with a "freakish" physique: at 6 foot 3, 235 pounds, with 2% body fat, he ran 40 yards in 4.39 seconds, had a 40.5-inch vertical leap, and bench pressed 440 pounds. Carter himself attributed his prowess in large part to Mickey D's. "I eat a lot of McDonald's and fast foods . . . [al]most every day. I usually get three double cheeseburgers, medium fries, large tea, and six-piece McNuggets. I don't think eating healthy as far as eating salads and that stuff really works for me." Or the fact that Usain Bolt stated in his autobiography that he ate approximately one thousand Chicken McNuggets while in Beijing for the 2008 Olympics, where he won gold medals in both the 100- and 200-meter races and a third in the 4-by-100-meter relay? How does one explain examples such as these without acknowledging that food and food systems are complex subjects of inquiry?[50]

Yes, yes, I know that, generally speaking, too much fast food is strongly and positively correlated with "markers" of poor health such as obesity and high BMIs, and also with diabetes, elevated blood pressure, arteriosclerosis, incidence of strokes, and so on. But I also know that the fast food, which is "energy-dense," as nutritionists say, provides a considerable caloric bang for a buck. "Dollar meals" may not appeal much to affluent consumers, but to groups further down the food chain, as it were, they hit the spot. To be sure, too many dollar meals may lead to health problems down the line, but, in economics lingo, the "pure time preference" for poor people is the present. They have every right to apply heavy discount rates to the future, which may not be all that great in any case.[51]

As I said, things are complicated. And we haven't even mentioned the Farm Bill, decried if not despised by the food police, even though 80% of the outlays in the 2014 iteration (PL 113-79), which ran through 2018, went to nutrition-assistance programs of one kind or another, most notably SNAP, formerly known as food stamps.[52] As Sarah Ludington points out in her article (chapter 8), those are precisely the same reasons that economic libertarians and social conservatives also despise the Farm Bill. Nor have we mentioned the recent questions raised about—and powerful challenges made to—narratives regarding "food deserts" in America's inner cities.[53] Those are difficult subjects, too, but subjects better tackled another day.

NOTES

1. See Peter A. Coclanis, "Agriculture as History," *Historically Speaking* 4 (November 2002): 3–4.

2. Robert Vic, "The Portland Canal" (senior honors thesis, history, University of North Carolina at Chapel Hill, 2002), 54.

3. Louis A. Ferleger, "Agriculture's Last Stand: A Note on the Missing South," *Journal of the Historical Society* 12 (March 2012): 97–106; Leslie Horn, "Facebook Dropped $8.5M for FB.com," *PC Magazine.com*, 12 January 2011, http://www.pcmag.com/article2/0,2817,2375643,00.asp.

4. Karl Marx, *The Eighteenth Brumaire of Louis Napoleon* (New York: International Publishers, 1964; originally published in German in 1852), 15.

5. On the mixed, "mestizo" character of early American agriculture and the important contributions made by various groups, see, for example, Russell R. Menard, "Colonial America's Mestizo Agriculture," in *The Economy of Early America: Historical Perspectives and New Directions*, ed. Cathy Matson (University Park: Pennsylvania State University Press, 2006), 107–23.

6. This observation, named after nineteenth-century German statistician Ernst Engel, states that, after income rises above subsistence levels, the relative proportion spent on food rises more slowly than income. Data on demand and income elasticities for food around the world in the contemporary era can be found in Andrew Muhammad, James L. Seale Jr., Birgit Meade, and Anita Regmi, *International Evidence on Food Consumption Patterns: An Update Using 2005 International Comparison Program Data*, TB-1929, U.S. Department of Agriculture, Economic Research Service, March 2011, revised February 2013, http://www.ers.usda.gov/media/129561/tb1929.pdf.

7. My interpretation of U.S. agricultural history is based on a lifetime of studying the topic. For excellent surveys of American agriculture and its history, see, for example, R. Douglas Hurt, *American Agriculture: A Brief History*, rev. ed. (West Lafayette, Ind.: Purdue University Press, 2002); Bruce L. Gardner, *American Agriculture in the Twentieth Century: How It Flourished and What It Cost* (Cambridge, Mass.: Harvard University Press, 2002). For classic analytical treatments, see William M. Parker, "Agriculture," in *American Economic Growth: An Economist's History of the United States*, ed. Lance E. Davis et al. (New York: Harper & Row, 1972), 369–417; Willard W. Cochrane, *The Development of American Agriculture*, 2nd ed. (Minneapolis: University of Minnesota Press, 1993).

8. Hurt, *American Agriculture*; Gardner, *American Agriculture in the Twentieth Century*; Parker, "Agriculture"; and Cochrane, *The Development of American Agriculture*.

9. See U.S. Department of Agriculture, Economic Research Service, *Ag and Food Sectors and the Economy* (Washington, D.C.: USDA, 2015, last updated 5 May 2017), https://www.ers.usda.gov/data-products/ag-and-food-statistics-charting-the-essentials/ag-and-food-sectors-and-the-economy.aspx.

10. See Parker, "Agriculture," 372–75.

11. U.S. Department of Agriculture, Economic Research Service, Food Prices and Spending (Washington, D.C.: USDA, 2014, last updated 25 April 2017), https://www.ers.usda.gov/data-products/ag-and-food-statistics-charting-the-essentials/food-prices-and-spending/; U.S. Department of Agriculture, Economic Research Service, Food Expenditure Series, last updated 1 December 2014, Table 7, Table 97, http://www.ers.usda.gov/data-products/food-expenditures.aspx. Note that table 97 is titled "Percent of Consumer Expenditures Spent on Food, Alcoholic Beverages, and Tobacco That Were Consumed at Home, by Selected Countries, 2013."

12. For a brief introduction to ADLI, see United Nations Economic and Social Council, "The Agricultural Development Led Industrialization (ADLI) Strategy," Development Strategies That Work series (2010), http://webapps01.un.org/nvp/indpolicy.action?id=124, accessed 17 January 2015. On agriculture's role in the development of the U.S. manufacturing belt, see, for example, Brian Page and Richard Walker, "From Settlement to Fordism: The Agro-Industrial Revolution in the American Midwest," *Economic Geography* 67 (October 1991): 281–315.

13. See World Bank, *World Development Report 2008: Agriculture for Development* (Washington, D.C.: World Bank, 2007). For an introduction to the Bill and Melinda Gates Foundation's agricultural development programs, see http://www.gatesfoundation.org/What-We-Do/Global-Development/Agricultural-Development, accessed 17 January 2015.

14. On this broad usage, see, for example, Nathan Rosenberg, *Technology and American Economic Growth* (New York: Harper & Row, 1972), 1–24.

15. On the importance of revolutionary breakthroughs in agricultural history, see, for example, Peter A. Coclanis, "Two Cheers for Revolution: The Virtues of Regime Change in World Agriculture," *Historically Speaking* 10 (June 2009): 2–7.

16. Alfred D. Chandler Jr., *The Visible Hand: The Managerial Revolution in American Business* (Cambridge, Mass.: Belknap Press of Harvard University Press, 1977), 64–67. Note that Caitlin Rosenthal has recently argued that southern plantations were much more "modern" in terms of business strategy and practices than Chandler believed. See, for example, Rosenthal, "From Memory to Mastery: Accounting for Control in America, 1750–1880," *Enterprise and Society* 14 (December 2013): 732–48; Rosenthal (interviewed by Scott Berinato), "Plantations Practiced Modern Management," *Harvard Business Review* 91 (September 2013): 30–31; Rosenthal, "Slavery's Scientific Management: Masters and Managers," in *Slavery's Capitalism: A New History of American Economic Development*, ed. Sven Beckert and Seth Rothman (Philadelphia: University of Pennsylvania Press, 2016), 62–86.

17. On the evolution of the U.S. manufacturing "regime," see, for example, David A. Hounshell, *From the American System to Mass Production, 1800–1932: The Development of Manufacturing Technology in the United States* (Baltimore: Johns Hopkins University Press, 1984).

18. Eric Schlosser, *Fast Food Nation: The Dark Side of the All-American Meal* (Boston: Houghton Mifflin, 2001), esp. 149–90; Karl Taro Greenfeld, "Taco Bell and the Golden

Age of Drive-Thru," *Bloomberg Businessweek*, 5 May 2011, http://www.businessweek.com/magazine/content/11_20/b4228064581642.htm; Ilya Leybovich, "What Manufacturers Can Learn from Fast Food," *IMT: Industry Market Trends*, 24 May 2011, http://news.thomasnet.com/imt/2011/05/24/what-manufacturers-can-learn-from-fast-food. For a provocative new critique of factory meat production—focusing on Hormel Foods and Spam—see Ted Genoways, *The Chain: Farm, Factory, and the Fate of Our Food* (New York: HarperCollins, 2014).

19. U.S. Department of Agriculture, *2012 Census of Agriculture*, vol. 1, chapter 1, U.S. National Level Data, Table 64, Summary by Size of Farm: 2012, http://www.agcensus.usda.gov/Publications/2012/Full_Report/Volume_1,_Chapter_1_US/st99_1_001_001.pdf, accessed 25 November 2018; U.S. Department of Agriculture, *2012 Census of Agriculture*, vol. 1, chapter 1, State Level Data [North Carolina], Table 1: Historical Highlights: 2012 and Earlier Census Years, http://www.agcensus.usda.gov/Publications/2012/Full_Report/Volume_1,_Chapter_1_State_Level/North_Carolina/st37_1_001_001.pdf, accessed 18 January 2015. Note that by 2015 the average-size farm in North Carolina had grown to 170 acres. Note, too, that between 2011 and 2015 the number of farms in North Carolina fell from 50,800 to 48,800. See U.S. National Agricultural Statistics Service and the N.C. Department of Commerce and Consumer Services, *North Carolina Agricultural Statistics 2016* (Raleigh, 2016), 10.

20. U.S. Department of Agriculture, *2012 Census of Agriculture*, vol. 1, chapter 1, U.S. National Level Data, Table 9, Land in Farms, Harvested Cropland, and Irrigated Land, by Size of Farm: 2012 and 2007, http://www.agcensus.usda.gov/Publications/2012/Full_Report/Volume_1,_Chapter_1_US/st99_1_009_010.pdf, accessed 18 January 2015. Note that the U.S. Census of Agriculture is taken every five years, the last in 2012. Questionnaires for the 2017 census will be mailed out in December 2017, and the results for the 2017 census will begin to be released in February 2019. See U.S. Department of Agriculture, Census of Agriculture, https://www.agcensus.usda.gov/Help/FAQs/2017/, accessed 18 May 2017.

21. U.S. Department of Agriculture, *2012 Census Highlights, Farm Economics*, ACH12-2, May 2014, http://www.agcensus.usda.gov/Publications/2012/Online_Resources/Highlights/Farm_Economics/.

22. William Neuman, "Strong Exports Lift U.S. Agriculture Sector," *New York Times*, 1 September 2010, http://www.nytimes.com/2010/09/01/business/economy/01exports.html?_r=0.

23. U.S. Department of Agriculture, Foreign Agricultural Service, *Value of U.S. Agricultural Exports, 1990–2016*, https://www.fas.usda.gov/data/value-us-agricultural-exports-1990–2016, accessed 18 May 2017. Note that in 2014 the value of U.S. agricultural exports peaked at about $150 billion. The strong dollar, lower prices for commodities, and increased international competition all played roles in the modest U.S. decline since 2014.

24. U.S. National Agricultural Statistics Service and the N.C. Department of Commerce and Consumer Services, *North Carolina Agricultural Statistics 2016*, 9; U.S. National Agricultural Statistics Service and the N.C. Department of Agriculture and Consumer Services, *North Carolina Agricultural Statistics 2013* (Raleigh, 2013), 9; *North Carolina Agricultural Statistics 2011* (Raleigh, 2011), 11.

25. For a good overview of the Open Grounds Farm, see Edward Martin, "High-Yield Investment," *Business North Carolina*, 12 January 2012, http://businessnc.com/high-yield-investmentcategory/.

26. U.S. Department of Agriculture, *Organic Farming: Results from the 2014 Organic Survey* (ACH12-29/September 2015), 2, https://www.agcensus.usda.gov/Publications/2012/Online_Resources/Highlights/Organics/2014_Organic_Survey_Highlights.pdf, accessed 19 May 2017.

27. Robert Paarlberg, "Attention Whole Foods Shoppers," *Foreign Policy*, May/June 2010, http://foreignpolicy.com/2010/04/26/attention-whole-foods-shoppers/.

28. Of the differential in yields between conventional and organic farming, see, for example, Paul Maeder et al., "Soil Fertility and Biodiversity in Organic Farming," *Science* 296 (31 May 2002): 1694–97; Verena Seufert, Navin Ramankutty, and Jonathan A. Foley, "Comparing the Yields of Organic and Conventional Agriculture," *Nature* 485 (10 May 2012): 229–32; Jayson Lusk, *The Food Police: A Well-Fed Manifesto about the Politics of Your Plate* (New York: Crown Forum, 2013), 92–93, 209.

29. Alan D. Dangour, Karen Lock, Arabella Hayter, Andrea Aikenhead, Elizabeth Allen, and Ricardo Uauy, "Nutrition-Related Health Effects of Organic Foods: A Systematic Review," *American Journal of Clinical Nutrition* 92 (July 2010): 203–10, http://ajcn.nutrition.org/content/92/1/203.long; Paarlberg, "Attention Whole Foods Shoppers"; Crystal Smith-Spangler et al., "Are Organic Foods Safer or Healthier Than Conventional Alternatives? A Systematic Review," *Annals of Internal Medicine* 157 (4 September 2012): 348–66, http://annals.org/article.aspx?articleid=1355685. Note that even those studies arguing for the nutritional superiority of organic food generally find only modest nutritional differences between organics and nonorganics.

30. Paarlberg, "Attention Whole Foods Shoppers." See also Lusk, *The Food Police*, 81–99.

31. See Andrew Martin, "Walmart Promises Organic Food for Everyone," *Bloomberg Businessweek*, 6 November 2014, http://www.businessweek.com/articles/2014-11-06/wal-mart-promises-organic-food-for-everyone; Angel González, "Largest Organic Grocer Now Costco, Analysts Say," *Seattle Times*, 4 April 2016.

32. James E. McWilliams, "The Myth of Sustainable Meat," *New York Times*, 13 April 2012, A23. For a recent defense of meat production and its sustainability, see Nicolette Hahn Niman, *Defending Beef: A Case for Sustainable Meat Production* (White River Junction, Vt.: Chelsea Green, 2014). Note that McWilliams is quite critical of the foodie fetish for localism. See McWilliams, *Just Food: Where Locavores Get It Wrong and How We Can Truly Eat Responsibly* (New York: Little, Brown, 2009). For a very spirited critique of food localism and locavores, see Pierre Desrochers and Hiroko Shimizu, *The Locavore's Dilemma: In Praise of the 10,000-Mile Diet* (New York: Public Affairs, 2012).

33. Ben Worthen, "A Dozen Eggs for $8? Michael Pollan Explains the Math of Buying Local," *Wall Street Journal*, 5 August 2010. See also Virginia Postrell, "No Free Locavore Lunch," *Wall Street Journal*, 25–26 September 2010, C10; Ronald Bailey, "Chipotle Treats Customers Like Idiots," *Reason* 47 (August/September 2015): 16–17.

34. U.S. Department of Commerce, Census Bureau, *Income and Poverty in the United States: 2015*, by Bernadette D. Proctor, Jessica L. Semega, and Melissa A. Kollar, 13 September 2016, Report Number P60-256, https://www.census.gov/library/publications/2016/demo/p60-256.html.

35. See U.S. Department of Agriculture, Economic Research Service, "Percent of Income Spent on Food Falls as Income Rises," by Charlotte Tuttle and Annemarie Kuhns, *Amber Waves*, 6 September 2016, https://www.ers.usda.gov/amber-waves/2016/september/percent-of-income-spent-on-food-falls-as-income-rises/.

36. On "high modernism," see James C. Scott, *Seeing Like a State: How Certain Schemes to Improve the Human Condition Have Failed* (New Haven, Conn.: Yale University Press, 1998), 4–6 and passim.

37. Paarlberg, "Attention Whole Foods Shoppers." See also Manitra A. Rakotoarisoa, Massimo Iafrate, and Marianna Pacchali, *Why Has Africa Become a Net Food Importer? Explaining Africa Agricultural and Food Trade Deficits* (Rome: Food and Markets Division, Food and Agriculture Organization of the United Nations, 2011), http://www.fao.org/docrep/015/i2497e/i2497e00.pdf.

38. For a provocative argument proposing a fusion of organic farming and genetic engineering, see Pamela C. Ronald and Raoul W. Adamchak, *Tomorrow's Table: Organic Farming, Genetics, and the Future of Food* (New York: Oxford University Press, 2008).

39. Organisation for Economic Co-operation and Development (OECD), *Environmental Performance of Agriculture in OECD Countries since 1990* (Paris: OECD, 2008), http://www.oecd-ilibrary.org/agriculture-and-food/environmental-performance-of-agriculture-in-oecd-countries-since-1990_9789264040854-en; OECD, *OECD Compendium of Agri-environmental Indicators* (Paris: OECD, 2013), http://www.oecd-ilibrary.org/agriculture-and-food/oecd-compendium-of-agri-environmental-indicators_9789264186217-en.

40. Peter A. Coclanis, "Food Is Much Safer Than You Think," *Wall Street Journal*, 14 June 2011, A13.

41. On ways to reduce food waste, particularly postharvest loss, see, for example, Peter A. Coclanis, "Low-Hanging Fruit: The Fight for Food Security," *Le Monde Diplomatique* (English edition), 29 December 2015, http://mondediplo.com/outsidein/low-hanging-fruit-the-fight-for-food-security; Coclanis, "There Is a Simple Way to Improve the World's Food Systems," *Aeon*, 27 February 2017, https://aeon.co/ideas/there-is-a-simple-way-to-improve-the-worlds-food-systems.

42. On some of the advantages of GMO crops, see, for example, Marc Van Montagu, "The Irrational Fear of GM Food," *Wall Street Journal*, 23 October 2013, A15; Lusk, *The Food Police*, 101–13; Joel Mokyr, "What Today's Economic Gloomsayers Are Missing," *Wall Street Journal*, 8 August 2014.

43. See, for example, Ronald Bailey, "The End of Farming," *Reason* 47 (June 2015): 22–23; Nicholas Kristof, "The (Fake) Meat Revolution," *New York Times*, 19 September 2015. It is both ironic—and telling—that Tyson Foods (of all food companies) has recently taken ownership of 5% of Silicon Valley synthetic meat producer Beyond Meat, one of the leaders in this rising field. See Stephanie Strom, "Tyson Foods, a Meat Leader, Invests in Protein Alternatives," *New York Times*, 10 October 2016.

44. On the Haber-Bosch process (for synthetically fixing nitrogen in plants through the production/processing of ammonia), see, for example, Vaclav Smil, *Enriching the Earth: Fritz Haber, Carl Bosch, and the Transformation of World Food Production* (Cambridge, Mass.: MIT Press, 2000). Smil, a noted historian of technology, considers the Haber-Bosch process the most important technological innovation of the twentieth century. Again, on the potential union of GMOs and organic agriculture, see Ronald and Adamchak, *Tomorrow's Table*.

45. On Food Week at Duke in October 2010, see https://foodatduke.wordpress.com/, accessed 19 January 2015. On the "Shared Tables" symposium, 28–29 February 2012, see https://sharedtablessymp.wordpress.com/, accessed 19 January 2015.

46. For more on my perspective regarding agriculture, see Peter A. Coclanis, "Food Chains: The Burdens of the (Re)Past," *Agricultural History* 72 (Fall 1998): 661–74; Coclanis, "Breaking New Ground: From the History of Agriculture to the History of Food Systems," *Historical Methods* 38 (Winter 2005): 5–13; Coclanis, "Two Cheers for Revolution: The Virtues of Regime Change in World Agriculture," *Historically Speaking* 10 (June 2009): 2–7.

47. Jayson Lusk, Food Demand Survey, 15 January 2015, http://jaysonlusk.com/blog/2015/1/15/food-demand-survey-foods-january-2015; Brandon R. McFadden and Jayson L. Lusk, "What Consumers Don't Know about Genetically Modified Food and How That Affects Beliefs," *FASEB Journal: The Official Journal of the Federation of American Societies for Experimental Biology* (30 September 2016): 3091–96, http://www.fasebj.org/content/30/9/3091. Note that in the McFadden and Lusk survey, 32% of those surveyed did not think that vegetables contained DNA!

48. Tara Parker-Pope, "Hard Road for Those Who Seek Veganism," *New York Times*, 17 April 2012, D1; Frank Newport, "In U.S., 5% Consider Themselves Vegetarians," *Gallup, Well-Being*, 26 July 2012, http://www.gallup.com/poll/156215/consider-themselves-vegetarians.aspx.

49. Steve Bertoni, "Sweet Investment: How Dean Metropoulos Made Billions Saving the Hostess Twinkie," *Forbes*, 28 February 2017, https://www.forbes.com/sites/stevenbertoni/2017/02/28/sweet-investment-how-dean-metropoulos-made-billions-saving-the-hostess-twinkie/#62c36fbc1f50. For another valuable assessment, particularly regarding the declining position of workers at Hostess Brands, see Rick Wartzman, "What Hostess Brands' Return to the Stock Market Says about the State of U.S. Workers," *Fortune*, 25 July 2016, http://fortune.com/2016/07/25/hostess-twinkie-ipo/.

50. Rebecca Leung, "The Subway Diet," CBS News, *48 Hours*, 2 March 2004, http://www.cbsnews.com/news/the-subway-diet-02-03-2004/; "Twinkie Diet Helps Nutrition Professor Lose 27 Pounds," CNN.com, 8 November 2010, http://www.cnn.com/2010/HEALTH/11/08/twinkie.diet.professor/; Samantha Grossman, "Teacher Loses 37 Pounds after Three-Month McDonald's Diet," *NewsFeed*, TIME.com, 5 January 2014, http://newsfeed.time.com/2014/01/05/teacher-loses-37-pounds-after-three-month-mcdonalds-diet/; Bruce Feldman, "UNC's Carter Heads 'Freaks' List," *Bruce Feldman Blog*, ESPN Insider, 24 June 2010, http://insider.espn.go.com/ncf/blog/_/name/feldman_bruce/id/5322140/north-carolina-lb-bruce-carter-biggest-athletic-freak-game; Laurie Stampler, "Usain Bolt Ate 100 Chicken McNuggets a Day in Beijing and Somehow Won Three Gold Medals," *NewsFeed*, TIME.com, 4 November 2013, http://newsfeed.time.com/2013/11/04/olympic-gold-medalist-reveals-beijing-diet-of-1000-chicken-mcnuggets-in-10-days/.

51. See Peter A. Coclanis and Fitz Brundage, "Fast-Food Region: Cheap, 'Energy-Dense' Eats in a Poor, Unhealthy Part of the United States" (paper presented at conference "State of the Plate: Food and the Local/Global Nexus," University of North Carolina at Chapel Hill, 27 March 2015).

52. See Renee Johnson and Jim Monke, *What Is the Farm Bill?*, Congressional Record Service, CRS Report 7-5700, 8 February 2017, https://fas.org/sgp/crs/misc/RS22131.pdf. See also Peter A. Coclanis, "One Man's Pork Is Another Man's Bacon," *Durham Herald-Sun*, 20 September 2015, C4. Federal involvement in agriculture goes back a long way in the United States, as Sarah Ludington details in her essay in this volume.

53. Gina Kolata, "Studies Question the Pairing of Food Deserts and Obesity," *New York Times*, 18 April 2012; John McWhorter, "The Food Desert Myth," *New York Daily News*, 22 April 2012; Joe Cortright, "Where Are the Food Deserts?," *City Observatory*, 5 January 2015, http://cityobservatory.org/food-deserts/. Note that a 2009 systematic review of the literature on food deserts concluded that they *do* exist in *parts* of the country but that their patterning and effects are complex. See Julie Beaulac, Elizabeth Kristjansson, and Steven Cummins, "A Systematic Review of Food Deserts, 1966–2007," *Preventing Chronic Disease* 6 (July 2009): http://www.cdc.gov/pcd/issues/2009/jul/08_0163.htm; Penny Gordon-Larsen, "Food Availability/Convenience and Obesity," *Advances in Nutrition* 5 (November 2014): 809–17. For an excellent—realist—primer on today's food world, see Robert Paarlberg, *Food Politics: What Everyone Needs to Know*, 2nd ed. (New York: Oxford University Press, 2013).

3

Food Activism

A Critical History

STEVE STRIFFLER

Central not only to human survival but also to economic, cultural, social, and physical well-being, food has long been an important site of political conflict and engagement. Such activity has come in a wide range of forms, from workplace battles in restaurants, processing plants, homes, and fields to protests about contamination and chemicals, riots over prices and shortages, campaigns to end hunger, and efforts to create fair trade networks, communes, and organic farms. The motivation behind such actions has been similarly varied. Through a politics of food, people in the United States have sought to organize themselves, support farmers and workers, promote policy changes, expose injustice, embrace new lifestyles, opt out of society altogether, and even revolutionize the very ways in which we produce, trade, consume, and think about food.

Despite this considerable variation, activism around food has tended to follow two broad currents that are critical of the conventional food system, yet work from essentially distinct understandings of change, which in turn lead to different points of departure for political engagement. The first current is less concerned with building alternative food systems than with directly confronting the conventional regime. This would include efforts to organize farmworkers, processing plant labor, restaurant workers, and even farmers, most of which were not understood as "food activism" until relatively recently, but also a wide range of policy and regulatory initiatives led by consumers, politicians, activists, and others. Most of these efforts do not look to create a parallel system, but seek to reform industrial food by improving wages, working conditions, food safety, environmental impacts, and the position of small farmers vis-à-vis agribusiness. Even the more radical versions within this tradition, those which seek to fundamentally restructure the way we do food, work from the assumption that change will come less by developing alternative universes outside of conventional food regimes than by transforming, even upending, the mainstream food system itself.

The second tradition of "food activism" has focused on conceptualizing and developing alternative ways of doing food (in terms of production, transportation, marketing, trade, consumption, etc.). Much of the organic movement, including related projects and precursors associated with back-to-the-land initiatives, counterculture cuisine, vegetarianism, and communes, all fit within this tradition, but so too do many contemporary urban gardening endeavors, the local food movement, CSAs, and fair trade. The construction of alternatives such as organic or fair trade are seen by proponents as posing a challenge to conventional foodways because they educate and inspire while demonstrating that other ways of doing food are possible; because they allow people to partially or fully "opt out" of mainstream food systems; and/or because they are seen as having the potential over time to supplant industrial agriculture or conventional trade. This alternative current is central to much of what we typically think of as "food activism" today.

Although this latter tradition pursues change by creating alternatives to the conventional food system, whereas the former looks to transform the food regime from within, there is not only considerable overlap between them, but both currents understand the conventional food system as being some combination of immoral, unhealthy, unfair, and unsustainable. More than this, the two currents embody a larger paradox that defines the history of food activism more broadly: how is it possible, given all this activism, that so much and yet so little has changed with respect to the politics and practice of food?[1]

To be sure, activism around food has mattered. It has been part of a sea change in terms of our relationship with food. Organic, for example, is no longer easily ridiculed or dismissed, but is in fact very much part of the mainstream. Many of us now get more enjoyment from a wider variety of foods, spend a great deal of time reading, writing, and talking about what and how we eat, and otherwise have a lot of fun with food. Likewise, thanks both to activism and an explosion of popular writing on the topic, more of us think critically about where our food comes from, what environmental impacts it generates, and how farmers are being treated. We are even beginning to think more about workers, those who cultivate, process, and serve the food we consume, as well as those who have not benefited from many of the positive changes associated with food during the past several decades. Food activism has clearly played a role in propelling these shifts while leading larger numbers of people to the conclusion that there is something profoundly wrong with the conventional food system (on the explosion of interest in food, see Coclanis in this volume).

At the same time, the simple fact is that on some level not much has changed, and in a structural sense a lot has gotten worse during the very period that food activism has been at its most vibrant and visible. Political activity around food has taken place for more than a hundred years now, and "food activism" has intensified from the 1960s to the present. Nevertheless, agribusiness and large retailers—now bigger and more concentrated than ever[2]—are firmly in control of a food system that is profoundly unhealthy for many of the workers, farmers, consumers, environments, and animals involved. Food workers tend to be poorly paid and labor under difficult, even dangerous, conditions;[3] farmers often live in extremely precarious circumstances, overly exposed to fluctuating demand and corporate control;[4] consumers are frequently enmeshed in a system that provides poor access to nutritional food while actively encouraging the consumption of unhealthy food;[5] the environment is being devastated by agro-industry;[6] and animals are subject to needlessly cruel treatment under the industrial model.[7]

To be sure, those activists who have sought to develop alternative ways of thinking about and doing food, through organic, fair trade, farmers markets, CSAs, or other initiatives, have had considerable success at moving the needle of public opinion, changing the way we think about food, and have even realized some important gains in terms of how we do food. But by and large these efforts have had remarkably little impact on corporate control or the broader structure of the industrial food system. Likewise, despite occasional bright spots in farmworker organizing, food safety regulations, or environmental policy, those who have attempted to confront the conventional food system head-on have had relatively little success in wresting control from the exceptionally large companies—backed by policy makers and regulatory agencies—who are the primary architects of a system that puts profit before the well-being of workers, farmers, animals, consumers, and the environment.[8]

How is it possible that decades of varied and vibrant food activism have changed so much and yet so little? Or, to put it slightly differently, given the immense variety and vibrancy within "food activism," how does one distinguish between those projects that are laying the groundwork for a movement or politics that will usher in significant transformation and those that are leading nowhere? The answer to this latter question will, of course, only come with time, but a turn to history can help us understand what types of struggle have been tried before, to what effect, and how they are or are not related to more contemporary battles. This seems particularly important with respect to food activism, an arena of political activity whose energy, wide variation, sense of urgency, and penchant for experimentation can easily lead activists and scholars to the conclusion that whatever is happening now must be new and pathbreaking (when in fact it may be part of a longer tradition). Likewise, such activism often contains quasi-messianic claims that a particular strategy or set of practices will not only be politically meaningful, but completely transform the way we produce, trade, transport, and consume food, when in fact many strategies have a long track record of producing very little in the way of change.

This chapter traces the two currents of food activism outlined above by focusing first on workers in the food sector, whose political activity until recently was understood less as "food activism" than as part of a broader working-class struggle, and then by turning to efforts to construct alternative food systems, particularly those around organic and fair trade, which are the precursors to what is currently understood as "food activism." As we will see, "food" worker organizing, although never particularly powerful, declined over the course of the twentieth century not only in the simple sense of having less capacity to improve wages and working conditions, but also in that it would enter the broader arena of food activism in a weakened state, as a minority current within the broader surge of "food activism" during the 1990s and 2000s (whose origins can largely be found in organic and to a lesser extent fair trade).

More than this, what the chapter suggests is that the declining power of labor within the food sector since the 1950s and the emergence of a more consumer-focused "food activism" since the 1960s are not entirely unrelated. Although labor's decline can hardly be blamed on the rise of food activism, the reduced capacity of working people to confront capitalism head-on—to even pressure the state to enact policies that mitigate against the worst excesses of the market—has not only made working people's lives increasingly difficult, but has allowed powerful sectors to elevate the market, and market-related solutions, to near hegemonic status. This has not only had disastrous consequences for working people and reduced their capacity to influence state power. It has also meant that as progressives have lost the capacity to shape state polices, and the state itself has become seen as a problem instead of a possible solution, activists have turned increasingly to the market (and consumers) for solutions to what are fundamentally political problems requiring mobilizations designed to confront state and corporate power.

This tendency, to divorce activism from anything resembling a left politics, is endemic to much of food activism today and partially explains why an increased awareness about food, as well as the remarkable proliferation of food-related activism, has done little to change the basic structure of the food system. Or, to put it differently, because the conventional food system cannot be changed in isolation from capitalism, meaningful transformation will come less from "food activism," especially one oriented around consumers and markets, than from a left politics aimed at challenging the larger structures that sustain capitalism, a system that puts profits over people, environments, and animals.

FOOD WORKERS

Labor organizing within the food sector has been an uphill battle. Although workers on farms, in processing plants, and within food service have long struggled for justice, it has proven hard to sustain organizing drives or realize lasting gains. Despite high-profile struggles by farmworkers, a history of successful unionization within meat processing, and persistent efforts to organize food service workers, close to half of today's food workers live in poverty, the by-product of a food regime that is controlled by agribusiness and makes it very difficult for workers to organize or improve their working conditions.[9] Like much else associated with food production, the lives of food workers, and the path toward building successful organizations, have always been difficult (on the broader difficulties associated with food production and consumption, see Rachel Laudan's chapter in this volume).

In California, the historical epicenter of industrial agriculture, farmworkers—including Chinese, Japanese, and Mexicans—were striking in the fields by the late 1800s and early 1900s. By the Depression-era 1930s, communist-led unions were organizing workers in sufficient numbers to pose a significant threat to corporate agriculture. Such efforts, however, were undermined by internal divisions and fierce opposition from powerful, well-connected growers, and then undone by World War II and the influx of braceros[10] during the 1940s and 1950s. Symbolic of this disempowerment was the exclusion of agricultural workers from the National Labor Relations Act of 1935, a measure that effectively denied agricultural labor the collective bargaining rights granted to industrial workers.[11] As Peter Coclanis demonstrates in chapter 2, American agriculture was in many respects revolutionary and a driving force behind U.S. economic development more broadly, but it was to a large extent a profoundly unequal system defined by powerful companies and an exploited labor force.

By the mid-1960s, however, when progressives brought the bracero program to an end and the political winds shifted, renewed efforts to organize slowly gained steam, eventually coalescing around the United Farm Workers (UFW) in the 1960s and 1970s. Despite remarkable success, and an innovative strategy that combined organizing in the fields, antipesticide campaigns, a national boycott, and hunger strikes, the movement declined during and after the 1970s due to external pressures and internal divisions and missteps. More recent efforts, most notably by the Farm Labor Organizing Committee (FLOC) beginning in the 1970s (in the Midwest and South) and the Coalition of Immokalee Workers since the 1990s, have generated great enthusiasm, promoted innovative tactics, and achieved notable gains, but despite more than a century of organizing, farmworkers remain largely unorganized while working and living in abysmal conditions.[12] Nor, as Margaret Gray's important book demonstrates, can we assume that workers on smaller family/boutique farms will find their conditions or capacity to organize any better.[13]

Likewise, not long after Upton Sinclair's *The Jungle* (1905) generated Progressive Era reform for consumers[14] but delivered little relief to workers and their families, workers in meat processing took it upon themselves to change their conditions. At the turn of the century, or roughly the same time as Chinese, Japanese, and Mexicans were striking in the fields of California, diverse groups of European immigrants unionized the massive enterprises owned by powerful companies like Armour and Swift. Black workers joined both the labor force and the struggle in the 1920s and 1930s, and the United Packinghouse Workers of America emerged as a powerful force, ensuring that workers not only in Chicago but in stockyards and packinghouses across the country had decent wages and working conditions while enjoying relatively stable, middle-class lives.[15]

This union-rooted security began to unravel in the postwar period as meatpackers, following the lead of a rapidly industrializing poultry industry, began to relocate meat processing to rural areas that were both closer to the animals and far more difficult to organize. Wages and working conditions declined during the latter half of the twentieth century as stockyards and packing facilities left places like Chicago and Omaha and dispersed throughout rural America. Today, beef, pork, and poultry production have relatively low levels of unionization and are known more for low wages, undocumented workers, dangerous working conditions, and human rights abuses than as a source of middle-class stability. In fact, according to a Human Rights Watch report on the U.S. meat industries: "*Many workers who try to form trade unions and bargain collectively are spied on, harassed, pressured, threatened, suspended, fired, deported or otherwise victimized for their exercise of the right to freedom of association.*"[16]

Workers within food service have followed a broadly similar pattern, though the history of this diverse sector is less well documented.[17] Unions of waiters, cooks, and bartenders emerged in the late 1800s and early 1900s in most major cities, continued to organize as women and blacks entered the workforce, and remained some of the most powerful organizations within the Hotel Employees and Restaurant Employees International Union (HERE) and other food-focused unions through the 1970s. In fact, although largely erased from public consciousness, waiters enjoyed relatively high levels of unionization by the 1940 and 1950s and, along with it, improved wages and greater control over the workplace. They led fights for better pay, shorter hours, and workplace conditions.

The decades after World War II, however, which saw the endless expansion of demand for food outside the home, brought a decline of unionization within food service, and with it decreased bargaining power, declining real wages, long hours, and intensified work. Although bastions of unionization remained, the rapidly expanding food service workforce of the 1960s and 1970s remained largely outside the union fold, a feature that would only intensify in the latter decades of the century as women, people of color, and immigrants became a much larger share of the food service labor force (for a deeper look at gender and labor see Deutsch's chapter in this volume). The general backlash against unions, the industry's unreal expansion, the broader turn to part-time labor, and the increasing power of corporate chains within the restaurant industry would eliminate the few unions that still existed in the 1980s while ensuring that new ones would not get off the ground.

The broad contours of this story are not particularly different for cooks, bartenders, butchers, or grocery store workers, the latter who enjoy higher rates of unionization but have faced the increased concentration of food retail under the control of companies such as Walmart, Kroger, Costco, and Target. Notwithstanding some innovative and high-profile efforts to organize fast food and grocery store workers, it is still the case that around half of all food service workers earn poverty wages within an industry that has proven remarkably adept at resisting efforts to organize or improve conditions.

Although the protagonists, supporters, and outside observers of the above struggles have not until relatively recently understood these efforts as "food activism," a current within this tradition—starting with Upton Sinclair, continuing with Cesar Chavez and the UFW, and carrying into the present—has clearly informed the more recent rise in consciousness around food. The relative success of these labor-oriented struggles, as well as their continued presence within the public imagination, is tied to their ability to connect with consumers and broader publics, including tactics (e.g., boycotts) and an orientation that has continued to inform activism through the present day.[18] Indeed, when "food activism" emerged on the scene as an identifiable, if vague, set of projects in the 1990s and 2000s, its most visible strands were drawn to the consumer-related tactics pioneered by the United Farm Workers and others. The irony, however, is that this move to consumer-oriented food justice not only brought with it a turn away from the labor movement and workers' concerns but was often disconnected from social movements and political mobilization altogether. It is to these market and consumer-oriented movements that we now turn.

ORGANIC

Building on earlier efforts, organic farming emerged in a recognizably modern form after World War II as a relatively coherent set of practices built around production techniques and land management. These early initiatives were then expanded considerably by a larger constituency that was heavily rooted in counterculture currents and drawn to organic in the late 1960s and early 1970s. Some of these advocates came out of civil rights and antiwar struggles, and were looking for new political avenues as 1960s-era radicalism lost steam or failed to transform society as quickly or deeply as advocates had hoped. This frustration with radical politics, and the intransigence of the political system more broadly, stimulated the counterculture return to the land, fed the impulse to opt out of urban-industrial life altogether, and fueled a range of radicalisms that shared a growing unease with modern-industrial-consumer society. Rural communes exploded between 1965 and 1970 (and then disappeared almost as quickly), reflecting a genuine angst with urban-consumer society and a growing sense that industrial agriculture was disastrous on a number of levels.[19]

Despite a countercultural haziness, a partial foundation in therapeutic self-enhancement and consumerist self-protection, and the consistent presence of a strand that pushed a more limited agenda of better ecology and healthier food, the core of organic during the 1970s was driven by organic farmers and their allies who understood the embryonic movement as a critique and challenge to the established food system, especially in terms of farm practices, food processing, and locally scaled distribution. It was to be a decentralized system defined by small-scale farmers who directly market products that are minimally processed. In short, it was oppositional. Its implementation would require a completely different system of food production, processing, and distribution.

Well into the 1970s, however, organic was very small, essentially a fringe practice that was the object of ridicule from mainstream America. The real rise of organic, in the sense of its dramatic expansion, did not take place until the 1980s. When this mainstreaming happened, moreover, it was not controlled by movement diehards but was ultimately driven by a consumer demand fueled in part by a series of food scares, was focused on the safety and health of food (for consumers), and was largely met by conventional growers who converted to organic for business reasons. This happened quickly, with organic representing close to 1% of all food sales by the mid-1990s,[20] with over half of the value of organic production going to only 2% of the growers.[21] In other words, movement growers and associated co-ops and consumers—those who understood organic as part of a transformative project—were increasingly marginalized within an organic world that was expanding exponentially (yet remained a tiny percentage of the overall food industry).

The corporate capturing of organic, its institutionalization through trade organizations and certifying agencies, and the increasingly narrow definition of organic as a production standard for farmers (as opposed to a food safety standard for consumers, a truly alternative food system, or even a political movement) was well under way by the time the Organic Foods Production Act of 1990 and the 1996 Food Quality Protection Act were passed. The acts, in a sense, confirmed what was already happening (or had happened): namely, that organic would be narrowly defined and run as a large industry with big business at the helm. Organic food would reach millions, but the radical impulses of "the movement" would be thoroughly marginalized when they did not simply become marketing material for the latest advertising campaign. Debate did not end, but it was increasingly irrelevant to the overall direction of an industry that was run firmly within the logic of capitalism.

Despite its remarkable "success," organic is nonetheless broadly representative of market-oriented efforts to build alternative food systems more generally. Such projects, even one as unusually successful as organic, typically meet one of two fates: either they remain so small that they pose no real challenge to the conventional food system or they are co-opted by business interests in such a way that drains them of most, if not all, of their radical content. Organic experienced both outcomes. The question remains why. Why has the alternative path for pursuing change not fared better? A turn to fair trade, which focuses more on the social costs of food for small-scale farmers, will help shed some light on that question.

FAIR TRADE

Fair trade began around the same time as organic was taking off. As U.S. religious leaders and lay activists from a range of denominations began to travel after World War II, they were stunned by the devastation in postwar Europe and by the poverty in the Third World. As the crisis in Europe faded, the focus turned to the Third World with a fairly straightforward goal: eradicate poverty by helping communities in the global South gain access to markets in the United States and Europe. Early on, the intent was not so much to challenge the existing system of conventional trade as to create alternative trade networks by purchasing handicrafts directly from poor producers and then selling those products to conscientious consumers in the North. Solidarity with impoverished regions would take place by making global trade available to the world's poor. Trade would increase employment/income, bring development to the Third World, and educate northern consumers about global poverty. Poverty, in short, could be eliminated through creative solutions rooted in the market.[22]

During the 1960s and 1970s, proponents became more critical of conventional trade. Sales through "World Shops" by Alternative Trading Organizations (ATOs) such as Ten Thousand Villages were the primary fair trade vehicle for combating poverty, which itself was increasingly understood as the product of an inherently unequal system of trade. By creating an alternative system of marketing and selling goods, the fair trade movement sought to return a larger portion of the market price to poor producers by eliminating the infamous middleman while at the same time educating northerners and giving them the opportunity to become conscientious consumers. The idea was to slowly grow the movement, thereby benefiting a larger number of producers while showing everyone—including big business—that an alternative model was possible. The challenge to the traditional system of (unequal) trade was the very existence of a fair trade system.

Although the ATO model experienced considerable success, in the grand scheme of things it was not even a tiny drop within the broader bucket of global trade. More than this, the intensified globalization of production and consumption from the 1980s forward served to highlight both the need for fair trade as well as the limited impact it was having on global poverty. On the one hand, globalization made it clear that consumers in the North were not only purchasing more from producers in the South, but were benefiting from cheap food, clothes, and electronics (etc.) made in the Third World. The devastating impact of this relationship on the global South became ever more apparent to growing numbers of Northern consumers. On the other hand, by further contributing to, and exposing, the rapidly growing inequality between the North and South, globalization demonstrated that despite over thirty years of fair trade there had been little progress toward the ambitious goal of eradicating poverty through alternative trade networks. In fact, things had and have gotten considerably worse. According to a recent United Nations report, high-income countries generated more than half of the world's income despite having only 16% of its population, whereas low-income countries held just above 1% of global income despite having 72% of the world's population.[23]

As a result, beginning in the 1980s, some within the movement wanted to expand fair trade by pushing it beyond handicrafts and the small retail world of nonprofit shops and catalogs. Handicrafts worked well as educational tools, and because they were visibly distinct from mass-produced items they were easily identifiable and attractive to conscientious consumers. The problem, however, was that the fair trade movement could only sell so many colorful rugs, handmade baskets, and intricate tapestries to sympathetic consumers in the United States and Europe.

Similarly, the ATO model of purchasing products directly from the producer and then selling them directly to the consumer through small retail shops kept the movement relatively "pure" by ensuring that retailers shared the movement's underlying principles. At the same time, if the movement was to expand, the "World Shop" model was limited. Mainstream consumers did not frequent World Shops. Simply put, a wider range of everyday products produced by Third World communities would have to be sold in a larger number of venues to a broader group of consumers if the movement was to have a greater market presence.

It was in this context of growing inequality between the North and South, as well as a deepening crisis for Third World farmers, that sectors within the fair trade movement looked to expand through a new model based on certifying and labeling food commodities. Food products, once having been labeled as "fair trade," would be sold in conventional supermarkets and other mainstream outlets. For some, the fair trade labeling of coffee—the movement's original and still dominant commodity—in the 1980s represents the beginning of fair trade as we think of it today. ATO sales are still significant, but labeled foods dominate the movement and represent its most dynamic sector.

The labeling/certifying sector of fair trade is committed to the same broad goals as the ATO model, but pursues them through different institutional forms and with a businesslike focus on expanding sales. With the ATO model, the fairness in fair trade is ensured by the fact that both producers and consumers trust the political commitment of the ATOs. By contrast, the labeling-model of fair trade ensures fairness by creating standardized measurements of what exactly constitutes fair trade, thereby allowing it to expand into greater numbers of stores and commodities. By specifying that a product is "fair trade," the label assures the consumer that this cup of coffee, bunch of bananas, or piece of chocolate benefits producers in certain ways regardless of where it is sold.

Although the model based on labeling and increased market share has largely won the day, debates among fair trade advocates continue, and follow a pattern that is remarkably similar to ongoing discussions surrounding organic. Is the goal to simply increase market share, with the idea of bringing a perhaps compromised, and narrowed, version of fair trade and organic to as many people as possible? And can this be done without ultimately being absorbed, or too severely deformed, by the market? Or is the point to empower workers and/or transform the unequal nature of trade relations? Or both? What types of changes will expand producers' access to the market and further their integration into the very market that the movement is working against, while at the same time deepening the movement's core values, empowering producer organizations, improving the economic situation of small producers, and transforming the conventional system of trade? More concretely, what does it mean when Nestlé is granted use of the fair trade seal, or when Starbucks starts to sell fair trade coffee? Is it a sign that fair trade has made it, and is now able to help many more small producers, or does it signal that fair trade has been emptied of all meaningful political content?

Most debates about agri-food alternatives tend to revolve around two interrelated sets of questions regarding (a) the depth of the challenge to conventional systems that a particular alternative represents and (b) the capacity to "scale up" and become something more than a niche. These debates are not unimportant, nor unrelated since the distinction is often between a vision that favors growth at any cost versus one that balances growth with some commitment to core principles (of equality, fairness, smallness, etc.).

The decision, for example, over whether to struggle for a narrow or a more expansive definition of organic is ultimately a battle over what defines "success," which in turn determines the kind of vision one is working toward. Is it enough to remove chemicals from the production process, or should organic also empower small farmers, help workers, and protect consumers? How one answers this question shapes strategies, defines allies and enemies, and determines the possibilities for expansion.

Indeed, the success of agribusiness in defining organic in relatively narrow terms ultimately ensured that it would be a source of immense profit and growth. This also meant that organic's "success" in reaching relatively large numbers of consumers would not represent a significant change or challenge to the conventional food system. Likewise, the fact that the narrow version of fair trade won the day allowed for large corporations—as producers, traders, and retailers—to enter the game and (minimally) expand the reach of (a handful of) fair trade commodities. It also meant that what came to constitute fair trade would neither generate extensive benefits for large numbers of producers nor challenge the global trade regime.

At the same time, without dismissing the importance of these debates and the struggles around them, they miss the point on at least two levels. First, the fact that the "narrow," more limited versions of organic and fair trade have consistently carried the day is not coincidental but indicative of a broader balance of power that ensures only watered-down alternatives will survive. It is extremely difficult to create alternative space within a larger capitalist system, but it is virtually impossible with something as central to the economy as food, where powerful interests fiercely protect the status quo and where emerging alternatives have to compete with an already-established food system dominated by some of the largest corporations in the world. This is partly why both organic and local food initiatives have had difficulty expanding beyond a certain point or reaching more diverse constituencies. It is also why efforts to expand necessarily require that the alternative, in order to be successful, has to look more and more like the conventional system it is trying to replace.

Moreover, it is not simply that alternative systems have to compete with conventional ones, but that they must do so within broader systems of banking, trade, marketing, production, and transportation that are hostile to alternative logics and are thoroughly supported by state power. The idea that one can (on a relatively large scale) produce, process, market, distribute, and consume a single (important) commodity, or sector of commodities such as food, in a way that is equitable and sustainable within a system in which every other commodity operates around a logic that is profit-driven, hypercompetitive, and destructive, is problematic. The "food system" cannot be changed in isolation because food is thoroughly embedded within capitalism. Consequently, what is required is less a "food activism" than a working-class movement designed to challenge state and corporate power more generally.

Second, there tends to be a serious overestimation of the extent to which market-based alternatives pose a threat to conventional markets, or capitalism more broadly. As Jaffee and Howard note, "They may pose a threat to status quo means of accumulation, but not necessarily to the process of accumulation,"[24] which in turns speaks to their susceptibility for being co-opted.

Put simply, the fact that market-based alternatives tend to get watered down and pose much less of a threat than advocates often suggest is due partly to the fact that market-based activism does not in any way resemble a left politics, rooted in political organizing, that is designed to shape not only public opinion, but influence public policy while confronting corporate and state power. That most food activism does not do this is both ironic and problematic, since this is precisely what is needed to change the food system. In this sense, although it is not insignificant whether organic or fair trade gets "co-opted" or "captured" by corporations and regulators or whether such co-optation or capturing is in fact inevitable, the broader point is that market-based forms of antipolitics are not equipped to bring about meaningful political change in the first place (despite often-enthusiastic claims to the contrary). This is significant given that these problems are, at their core, political.

CONCLUSION

One of the great virtues and contributions of the left has been, and has to be, its ability to articulate alternative visions of the world. In this respect, food has been exceptionally good to chew on. Because it is so central to life, and something that all human beings need both physically and culturally, it is also a particularly powerful tool for demonstrating that the "free market" should not be allowed to determine the nature of, and access to, everything we produce and consume. If we allow the market to run wild, and profit to be the driving force, then we might very well end up with a food system like the one we have now, which is unhealthy for the majority of workers, farmers, consumers, animals, and environments involved.

On this educational level, the food movement has had considerable success in demonstrating that the conventional food system is deeply problematic, unhealthy, and exploitative, that it is organized around profit and not human need. This success in terms of developing a critical, public, awareness about food partially explains the explosion of food activism since the 1990s, including not only ongoing efforts to organize workers and develop organic and fair trade, but the local food movement, CSAs, farmers markets, food justice, food security, food sovereignty, anti-GM initiatives, slow food, guerrilla gardening, urban gardening, school-related initiatives, freeganism, food banks, animal rights, and so on.

Yet, as useful as food is as an educational tool, as a way of getting people to think critically about the conventional food system, it is important to remember that food is really a metaphor. Put simply, it is not the conventional food system that is the problem. Instead, the problem is capitalism, a system whose very logic prioritizes profit over people and relies on a continually expanding economy that devours ever-greater quantities of natural resources. This recognition is absolutely crucial for thinking about where and how to channel political energy. In this respect, as important as it is to know where we might want to end up—what the alternative might look like down the road—it is equally important to think quite seriously about the strategies for getting there. This type of political discussion is remarkably undeveloped within most food activist circles, which is part of the explanation as to why there has been so much change and yet so little. The near-absence of alternative visions is partly due to the ascendancy of neoliberalism, which has not only placed limits on the alternatives we are able to envision but has also severely circumscribed the kinds of politics that many activists now consider to be part of the progressive tool kit. Activists are often quite willing to forgo, sacrifice, or simply not see the possibility for large-scale change, especially as access to farmers markets, organic food, and locavore restaurants continues to expand. In other words, they tend to confuse the changes in their own food-buying options with broad systemic change.

If the conventional food system existed in isolation, or was simply an anomaly in an otherwise equitable economic system, then trying to construct (more or less) independent alternatives might make more strategic sense as a path for change. But this is not the case, and the focus on creating alternatives is problematic not only because they are either embedded in the market and/or simply not producing much in the way of meaningful change, but also because they often embrace an antipolitics—rooted more in consumers than citizens—that move us further away from precisely what we need, the construction of a left that is capable of advancing and mobilizing working power in order to bring about meaningful transformation.

NOTES

1. Warren Belasco asks the same question in the preface to the updated second edition of his classic, *Appetite for Change: How the Counterculture Took on the Food Industry* (Ithaca, N.Y.: Cornell University Press, 2007), xi.

2. Currently, four companies control over 80% of the beef industry, 66% of the hog industry, almost 60% of the poultry industry, over 90% of soybeans, 80% of corn, and 50% of the seed industry. "I Keep Hearing about 'Concentration' in Farming. What Does That Mean and How Does It Affect Me?," Farmaid.org, January 2010, https://www.farmaid.org/issues/corporate-power/i-keep-hearing-about-concentration-in-farming-what-does-that-mean-and-how-does-it-affect-me/. Also see *Global Research* for reliable data on corporate concentration in the food sector: http://www.globalresearch.ca/big-corporations-have-an-overwhelming-amount-of-power-over-our-food-supply/5391615, accessed 24 November 2018. Also see Christopher Leonard, *The Meat Racket: The Secret Takeover of America's Food Business* (New York: Simon and Schuster, 2015).

3. Farmworkers have the lowest annual incomes of any sector of U.S. workers; some earn less than minimum wage, and close to one-third live in poverty. When accounting for inflation, farmworker wages have actually declined by about 20% during the past two decades. For basic statistics on farmworkers, see the National Farm Worker Ministry, http://nfwm.org/, accessed 24 November 2018; Student Action with Farmworkers, http://saf-unite.org/, accessed 24 November 2018; and the U.S. Department of Agriculture, http://www.ers.usda.gov/topics/farm-economy/farm-labor.aspx, accessed 24 November 2018. For their part, nearly half of all restaurant workers live at or near poverty, and they have more than twice the poverty rate of the overall labor force. The federal tipped minimum wage has been stuck at $2.13 for over twenty years, and the real wages for tipped workers are lower today than in 1966. For up-to-date information on service workers see the Restaurant Opportunities Center, which has produced a series of excellent reports, and the Economic Policy Institute: http://rocunited.org/, accessed 24 November 2018; http://www.epi.org/publication/waiting-for-change-tipped-minimum-wage/, accessed 24 November 2018. Real wages in meatpacking, which used to have high levels of unionization, have plummeted, from nearly $20 an hour in 1980 to about $13 an hour in 2013. UC Davis's *Migration Dialogue* has nice coverage around issues of meat and migration: http://migration.ucdavis.edu/rmn//more.php?id=1778, accessed 24 November 2018.

4. On the precariousness of U.S. small farms over time and the increasing control of corporate agriculture over those farms, see the Pew Commission, which has issued a series of excellent reports: http://www.pcifapia.org/media/index.html, accessed 24 November 2018.

5. A good place for understanding how bad it has gotten for consumers is Marion Nestle's *Food Politics: How the Food Industry Influences Nutrition and Health* (Berkeley: University of California Press, 2013). See also Michele Simon's *Appetite for Profit: How the Food Industry Undermines Our Health and How to Fight Back* (New York: Nation Books, 2006).

6. A recent report by the Environment American Research and Policy Center notes that as a direct result of growing corporate concentration within agribusiness the United States now has over 100,000 miles of rivers and streams that are too polluted for basic activities such as swimming and fishing: http://www.environmentamerica.org/sites/environment/files/reports/EnvAm_Ag_v6_print.pdf, accessed 24 November 2018. For current information about the impact of agribusiness on the environment see the websites of the Pew Commission and the Union of Concerned Scientists: http://www.ucsusa.org/, accessed 24 November 2018.

7. See Daniel Imhoff, ed., *The CAFO Reader: The Tragedy of Animal Industrial Factories* (New York: Earth Aware Editions, 2010).

8. For a broad overview of the negative impacts of increasing corporate control over the food system, see Karl Weber, *Food, Inc.: How Industrial Food Is Making Us Sicker, Fatter, and Poorer—and What You Can Do about It* (New York: Public Affairs, 2009).

9. On recent struggles to unionize, see David Bacon's "Grapes of Wrath: California Farmworkers Fight to Unionize," *Al Jazeera America*, 16 January 2015, http://america.aljazeera.com/articles/2015/1/16/grapes-of-wrath-cafarmworkersfighttounionize.html.

10. The Bracero Program was an agreement between the United States and Mexico to allow for the importation of temporary contract workers, or braceros, to work in U.S. agriculture. Started in 1942 and promoted as a way to address a wartime labor shortage, the program continued for more than two decades due to pressure from agribusiness.

11. The classic in the field is Carey McWilliams, *Factories in the Field: The Study of Migratory Farm Labor in California* (1939; reprint, Berkeley: University of California Press, 2000).

12. On the UFW see Matt Garcia, *From the Jaws of Victory: The Triumph and Tragedy of Cesar Chavez and the Farm Worker Movement* (Berkeley: University of California Press, 2012). On the early history of FLOC see W. K. Barger and Ernesto M. Reza, *The Farm Labor Movement in the Midwest: Social Change and Adaptation among Migrant Farmworkers* (Austin: University of Texas Press, 1994).

13. Margaret Gray, *Labor and the Locavore: The Making of a Comprehensive Food Ethic* (Berkeley: University of California Press, 2013).

14. In the form of the Meat Inspection Act of 1906 and Pure Food and Drug Act of 1906.

15. See Rick Halpern, *Black and White Workers in Chicago Packinghouses, 1904–54* (Champaign: University of Illinois Press, 1997); Roger Horowitz, *Negro and White, United and Fight! A Social History of Industrial Unionism in Meatpacking, 1930–1990* (Champaign: University of Illinois Press, 1997).

16. Human Rights Watch Report, "Blood, Sweat and Fear: Workers' Rights in U.S. Meat and Poultry Plants" (2004), http://www.hrw.org/reports/2005/usa0105/, accessed 24 November 2018.

17. The history for this section on service workers is drawn from Dorothy Sue Cobble's pathbreaking book, *Dishing It Out: Waitresses and Their Unions in the Twentieth Century* (Urbana: University of Illinois Press, 1991).

18. Matt Garcia develops this important point in the introduction to his *From the Jaws of Victory: The Triumph and Tragedy of Cesar Chavez and the Farm Worker Movement* (Berkeley: University of California Press, 2012).

19. This section on the organic movement draws primarily from Julie Guthman, *Agrarian Dreams: The Paradox of Organic Farming in California* (Berkeley: University of California Press, 2014), as well as Belasco, *Appetite for Change*, and Daniel Jaffee and Philip

H. Howard, "Corporate Cooptation of Organic and Fair Trade Standards," *Agricultural Human Values* 27, no. 4 (2010): 387–99.

20. Jaffee and Howard, "Corporate Cooptation," 2.

21. Guthman, *Agrarian Dreams*, 43.

22. This section on fair trade draws largely from Laura Raynolds, Douglas Murray, and John Wilkinson, *Fair Trade: The Challenges of Transforming Globalization* (London: Routledge, 2007), especially chapter 2 (Laura Raynolds and Michael Long, "Fair/Alternative Trade: Historical and Empirical Dimensions"), as well as Daniel Jaffee, *Brewing Justice: Fair Trade Coffee, Sustainability, and Survival* (Berkeley: University of California Press, 2007).

23. United Nations Department of Economic and Social Affairs, *Inequality Matters: Report on the World Social Situation 2013* (New York: United Nations, 2013), http://www.un.org/esa/socdev/documents/reports/InequalityMatters.pdf.

24. Jaffee and Howard, "Corporate Cooptation," 2.

Section II
Choosing Food

Section II considers long-standing questions important to both philosophers and social scientists. For philosophers, the question is, do some things taste objectively better than others, and if so, how do we measure "better"? For social scientists, the question is, are the choices we make when eating and drinking largely determined by family status, education, and income, that is, our social class? Moreover, is information about what is healthy based solely on science, or is it too a product of class values? Many consumers answer these questions by arguing that the best food is that which tastes best to the individual doing the tasting. Others, though, argue that some things clearly taste better than others and that their judgment is objectively true. Nevertheless, members of both these camps are just as likely to use taste as a way to judge the moral value or intelligence of others. Likewise, those who believe that individuals have complete free will, and thus complete responsibility over their own choices, argue that those who eat unhealthy food have no one to blame but themselves, whereas those who eat healthy food should be praised for their ability to treat themselves with respect. It follows that those who do not have self-respect can hardly demand respect from others, while those who do, can.

This debate matters because having good taste and eating healthy food is used all the time as a way to justify that those at the top, middle, and bottom of society deserve their fate. But what if taste and eating healthily have nothing to do with individual morality but instead reflect as well as concretize a social hierarchy that allows some people to dominate others? Should we then throw out all aesthetic judgment and healthy eating advice as a form of class oppression? That seems both extreme and foolish. If nothing else, making aesthetic judgments is fun, and it's not even clear that we can help ourselves anyway. If you taste a plastic-wrapped tomato in winter and a fresh-picked one in summer, your palate and brain can't help but record that one tastes better than the other. And while one diet never fits all, we recognize in ourselves that some foods really do produce healthier and happier bodies than others. Thus, one way to reframe these debates might be first to acknowledge the paradox that taste can be both socially and aesthetically constructed and that nutrition science contains a great deal of objective truth, even if that truth is embedded within subjective cultural values, and those values are usually determined by the elites. Once we do that, the questions in this debate become less about judging others and more about asking ourselves why we have the taste that we do and, just as important, why others might have the taste that they have.

4

Can "Taste" Be Separated from Social Class?

S. MARGOT FINN

In 2009, a *New York Times* headline asked: "Is a Food Revolution Now in Season?" Many signs seemed to be pointing to yes: the Obamas had planted a vegetable garden on the White House lawn, organic products had made the leap from Whole Foods to Walmart, and celebrities like Oprah and Gwyneth Paltrow had become champions for food variously described as *healthy, local, organic, sustainable, authentic, clean, slow, real,* or simply *good.*[1] The turn away from industrial, modern processed food that Rachel Laudan describes in chapter 13 was hailed as a burgeoning social movement that includes everyone from public health officials seeking to halt the spread of obesity to chefs who say local ingredients taste better to environmental activists who want to reduce pesticide use and slow climate change. However, the article offered at least one reason for doubt: as of 2008, organics represented only 3% of the U.S. food and beverage industry. Gary Hirschberg, CEO of the organic dairy brand Stonyfield Farm, noted that this made the entire organic sector essentially equivalent to a "rounding error."[2] As Peter Coclanis mentions in chapter 2, by 2012 it had increased to a mere 4%, where it remained as of 2014.[3] Apparently, the vast majority of Americans remain unable or unwilling to buy into the *good* food hype.[4]

This contradiction between the widespread visibility of the "food revolution" and its apparent failure to affect how most Americans actually shop and eat has prompted much debate about elitism in the food movement. Is eating *good* food elitist? Again, many signs point to yes. In general, the foods and practices celebrated by the movement require more money, time, and effort than the conventional alternatives. Shopping at specialty markets, adopting a vegan or Paleo diet, and dining at farm-to-table restaurants aren't just ways of trying to eat better; they also serve as potent status symbols. Nonetheless, as the word "elitist" functions as a kind of slur in contemporary American discourse, many of the leaders of the food revolution have been eager to defend against it.

The most common response to charge of elitism is to admit that yes, the movement so far has been primarily restricted to a relatively wealthy few due to the higher cost of *good* food; however, that doesn't make it elitist. What's really elitist is the system that prevents most people from having enough money, time, or knowledge to eat those things regularly. For example, in an interview about the backlash against the food movement, Michael Pollan told Ian Brown of the *Globe and Mail,* "To damn a political and social movement because the people who started it are well-to-do seems to me not all that damning. . . . The reason that good food is more expensive than cheap food is part of the issue we're trying to confront."[5] In other words, the movement may reflect the inequalities that shape our society, but it's really working to fight those inequalities. If anything, it is populist, not elitist. Alice Waters has also admitted that good food tends to be more expensive, but she insists that it's worth it: "We're talking about health, we're talking about the planet, we're talking about the people who are supporting the land. . . . Make a sacrifice on the cell phone or the third pair of Nike shoes."[6] The implication seems to be that even the poor in America can afford some luxuries and ought to make *good* food one of them.

Others dispute the assumption that *good* food is more expensive. In 2007, Nina Planck, the founder of the London Farmer's Market and New York City's Greenmarket, told *Plenty* magazine: "Organic food and whole food—what I call traditional food—is frugal. Buy a whole chicken. It serves four people twice—the second time as soup. Buy fresh, local produce in season and canned wild Alaskan salmon. . . . In the center aisles are processed, nutrient-poor, high-profit-margin foods. That's what will eat up your budget."[7] By inverting the typical representation of junk food as cheap and organic food as expensive, she also recasts the food movement as populist, not just in its ambitions but also in the sense of already being accessible to anyone who knows how to shop strategically. In a similar vein, *New York Times* columnist Mark Bittman has repeatedly claimed that cooking what he calls "real" food like a roast chicken and salad for dinner is cheaper (and tastier) than feeding a whole family at McDonald's.[8] They both imply that if there's a barrier to broader participation, it's not affordability but ignorance.

A third tactic employed by the food movement's defenders is dismissing the debate about elitism as misleading and possibly invented by public relations firms and lobbyists who represent the food industry and big agricultural interests. In a 2011 *Washington Post* opinion column titled "Why Being a Foodie Isn't Elitist," Eric Schlosser claimed that the elitist epithet was a "form of misdirection" used by groups like the American Farm Bureau Federation to "evade serious debate" about things like the current agricultural subsidy system that Sarah Ludington discusses in chapter 8 and that disproportionately benefits the wealthiest 10% of farmers. Schlosser writes: "It gets the elitism charge precisely backward. America's current system of food production—overly centralized and industrialized, overly controlled by a handful of companies, overly reliant on monocultures, pesticides, chemical fertilizers, chemical additives, genetically modified organisms, factory farms, government subsidies and fossil fuels—is profoundly undemocratic. It is one more sign of how the few now rule the many."[9] Schlosser's portrayal of the food movement as a democratic revolt against a real elite represented by the food industry echoes Waters's plea to support farmers (presumably small-scale ones) and Planck's reference to "high-profit-margin" processed foods in the center aisles, but he sidesteps the question of whether the many are included in the resistance. His attitude is echoed in the documentary *Food, Inc.* by farmer Joel Salatin, who responds to questions about whether it's elitist to expect people to pay the prices he charges for free-range eggs and meat by declaring the issue "specious."

These attempts to defuse or dismiss the debate about elitism haven't worked. It persists like a proverbial thorn in the food movement's side, or perhaps its Achilles' heel. Part of what may make the accusation of elitism so difficult to shake is the frustrating elusiveness of the issues at stake. Even the food movement's defenders can't seem to agree about whether *good* food is actually more expensive or whether the people who buy *good* food represent a rarefied few or the democratic masses. The unresolved questions about class in the food movement can be broken down into roughly three categories.

The first is essentially demographic. *Who* is taking part in this so-called food revolution? Is the movement exclusively (or at least disproportionately) rich, white, urban, liberal, educated, or otherwise unrepresentative? Or do privileged people merely make up the public face of the movement? Are poor and working-class, nonwhite, rural, conservative, uncredentialed, and otherwise marginalized people equally invested in *good* food, but perhaps acting on their convictions in ways that may go unrecognized?

A second set of questions concerns the motives of the participants. *Why* are some people driven to participate in the food revolution? Is eating "better" primarily an issue of cost and availability, as Pollan and Waters would have it, or are the people who buy and promote the foods associated with the revolution just better informed, as Planck and Bittman suggest? The issue of motive implicitly raises questions about the foods consecrated as "better" by the food revolution. Are they really healthier, better-tasting, and more environmentally friendly? According to whom and based on what evidence? Is it possible that the organic, local, and artisanal aren't really better at all and the food revolution is just another rarefied taste culture that serves primarily to distinguish those who can afford to engage in it from those who can't?

A third set of questions involves the effects of the movement. *So what* if participation in the food revolution is a mark of class privilege? Does it nonetheless work to fight inequality and injustice? Is buying organic or local produce a meaningful form of protest against unsustainable, unsafe, or unfair conditions in the food industry? Or does it actually make inequality and injustice worse by stigmatizing the tastes and eating habits of the masses, thereby reinforcing pernicious social hierarchies? Can a movement that excludes the working class and poor, deliberately or not, nonetheless work on their behalf? Or has the movement focused on issues like animal welfare and the environment, sometimes seen as bourgeois preoccupations, at the expense of issues like hunger, wage stagnation, and labor conditions?

Even if food industry advocates have encouraged the debate about elitism in the food movement as some kind of public relations subterfuge, they did not invent the status hierarchies and anxieties that give these questions broad cultural resonance. You don't have to be a shill for the food industry to wonder whether a social movement based largely on shopping differently (or "voting with your fork") can really be inclusive or create meaningful change. Furthermore, popular stereotypes about who shops at farmers markets, makes their own kombucha, and drinks shade-grown pour-over coffee reflect the reality that how we eat communicates important information about our identities, including class background and status. Class shapes not only what kinds of food people have access to, but also their desires, their beliefs about food, and their sense of taste. Of course, taste is shaped by many factors other than class, several of which are illuminated by Charles Ludington's analysis of wine and beer preferences in England, Scotland, and the United States. As his examples demonstrate, tastes can reflect engrained cultural practices, political allegiances, and beliefs about gender and authenticity, all of which are interrelated with class. Accounting for particular tastes, as Ludington does, requires attention to a multiplicity of social forces and identities. However, as I think the anxiety about elitism in the food revolution attests, taste is so profoundly shaped by class that the relationship merits special attention.

Precisely what the food you eat communicates about your class status is historically and geographically variable. Quinoa might indicate upper-class worldliness and health consciousness in America today, but it was a staple food of the masses in Peru for centuries until growing demand in the global North made it more expensive than wheat bread and pasta.[10] However, some patterns appear to be widely, if not universally, true. Foods that are scarce and require more resources to acquire or prepare (including time and labor) tend to be associated with higher status. As Ludington argues, people tend to prefer familiar tastes. Thus, they often have a special affinity for the foods they grew up eating, which is one reason that tastes tend to reflect one's class background. On the other hand, people also seek out novel foods, especially those that confer status. This quest for status is complicated by the risk status-seeking always carries of being declared a fraud, exemplified by the notion of "food snobbery."

At the very least, these patterns suggest that the class issues reflected in the debate about elitism should not be dismissed as "specious." If the demographics of the movement seem unrepresentative, it's worth trying to figure out whether that's true, and if so, why and how it might influence the movement's motives and effects. This chapter offers a brief exploration of the relationship between food and class in modern American capitalism that might help guide and inform attempts to answer those questions. First, I'll discuss how the philosophy of taste changed with the acknowledgment that people use things like food to jockey for competitive advantage in status hierarchies. Next, I'll discuss a few examples of how anxieties about class have shaped dietary reform movements and the social construction of *good* food. Finally, I'll trace the emergence and evolution of the idea of "food snobbery" to show how the rise of the modern class system and ongoing shifts in class structure affect the popular discourse about *good* food. The history of food and class does not offer any simple answer to the question "Is eating *good* food elitist?" but may help explain the question's tenacity and offer some insights for those seeking to understand how class shapes the contemporary food movement.

The idea that class influences taste is a fairly recent idea. The dominant aesthetic philosophy that emerged in the European Enlightenment presumed that taste was essentially universal. A classic example of this view is David Hume's 1757 essay "Of the Standard of Taste." As Robert Valgenti elaborates in his history of food in philosophy (chapter 11), Hume acknowledges that people may *seem* to have subjective, idiosyncratic preferences, but he argues that ultimately we're all predisposed to enjoy the same things.

Our preferences vary because of differing capacities of discernment, or what he calls "delicacy."

By way of illustration, Hume recounts the story from *Don Quixote* in which two men who are supposed to be wine experts are asked to pass judgment on a cask of wine. One says it's good except for a faint taste of leather; the other says it's good except for a faint taste of iron. The rest of the company ridicules them, saying it just tastes good. But the experts are vindicated when they finish the wine and, at the very bottom of the cask, discover a key on a leather strap.[11] The rubes who liked the wine don't have *different* taste preferences than the experts—it's not that they like the taste of iron or leather—they just lack the ability to detect the defects caused by the key and its strap. Although he admits that some differences in discernment may be innate, Hume argues that most people develop delicacy in different aesthetic realms through repeated exposure, especially to excellence. Taste enough wine, particularly good wine, and eventually you too may become "sensible of every beauty and every blemish."[12] As Valgenti argues, this subsumes the idea of subjective taste preferences in universal standards that serve traditional hierarchies of power. After all, who but the elite have ever been able to taste enough *good* wine to develop the delicacy to judge it accurately?

Hume's view was largely echoed by Immanuel Kant in *The Critique of Judgment*. Kant argues that there's a common sense (*sensus communis*) of beauty and everyone would converge on the same aesthetic judgments if they all had enough education and leisure time to cultivate good reason, imagination, and perception.[13] Any differences in taste, then, are due to some people being deficient in those areas. It's probably not surprising that patrician intellectuals of the eighteenth century would think the carefully cultivated tastes of their own educated, leisured class were superior. However, the philosophy of universal and hierarchical taste seems a bit out of place in our more relativist, pluralist age, when academics hold conferences devoted to *Jersey Shore* and everyone from fast food workers to the president of the United States is expected to be conversant in the same pop music and must-see TV. Yet that's precisely what the notion of *good taste* represents.

On the one hand, many people today do believe that taste is at least somewhat subjective. They know that one person's favorite food might be repellent to someone else, even if they share the same background. On the other hand, many people also believe on some level that some people have *good taste* and others have *bad taste*, which amounts to a ranking of some people's preferences above others. And it's not as simple as everyone thinking their own taste is superior. While most people like to believe they personally have good taste, there's no perfect correspondence between someone's individual predilections and their sense of the broader social standards represented by *good taste.* For example, you might personally hate bananas and yet not think that anyone who happens to like them has *bad taste.* Or you might really enjoy some benighted food like American cheese, knowing your affinity is at odds with what counts as *good taste* in the dominant culture today. Nor is it as simple as good taste corresponding to the most popular things. As in Hume and Kant's day, *good taste* is often associated with the carefully cultivated preferences of privilege.

If Nietzsche was the turning point for food in philosophy, as Valgenti argues, Pierre Bourdieu serves that role in sociology. Bourdieu turned the theory of universal and hierarchical taste standards on its head in his landmark 1979 book *Distinction: A Social Critique of the Judgment of Taste.* He and his team of researchers gave 1,217 French people drawn from all different social classes an extensive questionnaire about their preferences, and they found no evidence of the universal tastes Hume and Kant had described. Instead, their respondents had distinctive "systems of dispositions (*habitus*) characteristic of the different classes and class fractions."[14] Poorer people tended to prefer "lowbrow" things, middle-class people liked "middlebrow" things, and rich people liked "highbrow" things. For example, the working class said they preferred their meals "plentiful and good," whereas the upper class said they liked the "original and exotic."[15] The issue was not that everyone shared the same preferences but had different capacities for detecting virtues or blemishes; people from different classes had completely contradictory ideas about what constituted a virtue or blemish. For the poor, exotic foods were undesirable; for the rich, they were desirable. Bourdieu argued that these differences were primarily the product of two things: socioeconomic background and the quest for status.

Many of the preferences recorded in the survey reflected fine divisions within classes: craftsmen had different tastes than small shopkeepers, who in turn were different from commercial employees although they were all similar in terms of income, and the same was true of engineers versus executives. However, the general pattern reflected in every area of culture Bourdieu's team asked about was that the lower classes reported preferences "identified as vulgar because they are both easy and common," while the higher classes favored "practices designated by their rarity as distinguished."[16] The latter are not elevated, according to Bourdieu, because they are objectively better or universally pleasing. On the contrary, "natural" enjoyment and all the things that most people find viscerally pleasing were typically seen as "lower, coarse, vulgar, venal, servile."[17] The reason the upper-class preferences work to set people apart, or *distinguish* them, is precisely because they involve knowing about and liking things that most people don't. In other words, rich people might not like exotic foods because they taste better in some objective way than more familiar foods, but because they're harder to get and thus set them apart from other people around them.

The social nature of taste doesn't mean that people always consciously choose what to like based on what they hope it will make people think about them. Although Bourdieu's respondents demonstrated that people are highly aware of the social significance of their tastes, their preferences were also shaped by their upbringing, education, and the social structures they inhabit. Some working-class people might reject exotic or delicate food as a conscious expression of their own class identity, deliberately shunning fancy foods they associate with rich, pretentious people. However, many are skeptical about or indifferent to exotic foods because those things simply don't reflect the tastes they've been exposed to and learned to like. Similarly, the richer people aren't necessarily just saying they like exotic foods because that's what they know is the highbrow thing to like, it may be that their upbringing and education has truly predisposed them to find those particular foods delicious. People's tastes are based partially on their deliberate attempts to perform a particular status or gain some competitive advantage and partially on the spontaneous, visceral attraction and revulsion they develop based on their upbringing and education.

One of Bourdieu's main contributions was revealing how the competition for the kinds of status that taste can confer, which he referred to as *cultural capital,* is rigged. The dominant social classes have disproportionate influence over the process of cultural legitimation, meaning they have more say over what counts as *good taste.* They tend to occupy the professions and control the institutions like schools, museums, and media industries that shape how people see the world. That gives them more control over the narratives people create and accept that give essentially arbitrary preferences social meaning. Thus, the dispositions of the elite get consecrated as *superior,* not just the predilections of wealth. Additionally, since the tastes of the wealthy are often defined by scarcity, elite cultural capital is inherently difficult to acquire.

However, Bourdieu's ideas have not achieved especially widespread currency outside academia. Proponents of the food revolution often assume that anyone with sufficient knowledge and exposure to "high-quality" food will develop the same taste preferences. They assume that the inherent superiority of the natural and real over the artificial, the whole over the refined and processed, and the small-batch over the industrial-scale is obvious to everyone, rather than seeing those preferences as a matter of socialization and ideology. However, it's likely that part of the reason these foods seem to appeal to the same demographics Bourdieu found to like "highbrow" tastes is due to the fact that they're distinguished from the mainstream and relatively scarce and thus communicate high status. Conversely, if poorer people are less likely to buy *good* food, even if it's sometimes less expensive than fast food or they willingly splurge on other luxuries, it may not be out of ignorance or lack of access but instead a different kind of socialization. The scarcity that makes organic, artisanal, and local foods seem desirable to the elite or aspirational middle class may make it seem pretentious and fussy to others. In some classes, consuming and especially serving people ample, familiar, reliable foods might be a better way of communicating care and belonging and of proudly asserting a nonelite identity.

Based on Bourdieu's theory of habitus and cultural capital, one might expect the elite to simply assume their preferences are universal and react with either confusion or scorn if members of the lower class evince different desires. However, in practice elite discourses about the eating habits of the poor and working class reveal conflicted attitudes. At times the lower classes have come under fire for eating differently than the elite, especially in the United States when the working class was composed largely of immigrants. At other times, dietary reformers set different expectations for the poor and middle class and criticized attempts by the former to eat foods that were seen as better suited to the latter. Finally, in some cases elites have applauded the poor for the thrift and virtue of their diets, usually while critiquing the middle and upper classes for their profligacy and excess. These disparate responses exemplify the range of ways the elite can construct and deploy cultural capital in the realm of food for their own social benefit.

The history of stigmatizing the food of the poor is probably as old as social classes themselves, but it may have intensified in the mid-eighteenth century along with the expansion of the middle class. Due to the precariousness of their position, the middle classes have always had a lot to gain by making very clear distinctions between their ways of speaking, dressing, and eating and the way the poor do all of those things, which they regularly portray as not just different but *wrong*. They poor are shunned for not being able to afford luxuries like meat and sugar and for purchasing them despite their penury. For example, the Irish dependence on potatoes as a staple food was often cited by British imperialists in the eighteenth century as a reason for their supposedly inferior character.[18] Around the same time, the British writer Arthur Young would observe disapprovingly that the poor at an almshouse he visited spent any money they had on tea and sugar when "it would be better expended in something else."[19]

At the turn of the twentieth century, the leaders of the emerging home economics movement in the United States expressed great concern about the exotic (and, they thought, unhealthy) foods eaten by recent immigrants. They were especially critical of the "excessive" consumption of coffee, alcohol, and spicy, pungent, and pickled foods, which they claimed would cause indigestion, stunted growth, excessive sexual appetites, impropriety, and disorderly behavior.[20] In the 1922 book *Foods of the Foreign Born in Relation to Health*, dietician Bertha Wood of the Boston Dispensary and Food Clinic (a public kitchen where immigrant mothers were taught how to poach eggs and make frugal, nourishing porridge) writes, "The Jewish children suffer from too many pickles, too few vegetables, and too little milk. . . . Excessive use of pickled foods destroys the taste for milder flavors, causes irritation, and renders assimilation more difficult."[21] Socialist writer John Spargo compared immigrant children's appetite for pickles to alcoholism in adults, which was often attributed at the time to cravings caused by an inadequate diet.[22] As spicy and pickled foods are now seen as nutritionally innocuous or even health-promoting and there was never any good evidence to the contrary, these beliefs are now recognized as a part of widespread classist and anti-immigrant attitudes.

In place of immigrants' preferred foods, home economists advocated creamed codfish, baked beans, corn chowders, Indian corn pudding, and oatmeal porridge, all things marked not only by their thrift but also their deliberate blandness and their association with New England, whose regional culinary traditions were in the process of being defined as American. This resolutely Puritan fare was promoted in public school cooking classes, women's colleges, instructional kitchens, home visits by charitable organizations, and the growing array of practical texts as a prescription for health, moral restraint, and social welfare. It's ironic, but revealing, that at the same time as home economists were using corn to Americanize immigrants, the food rations provided by the federal Indian Bureau specifically excluded it. According to historian Donna Gabaccia: "To prevent starvation, the federal Indian Bureau provided reservation food rations—and these typically did not include corn. . . . While domestic scientists saw corn-eating as a way to Americanize new immigrants, they seemed eager to wean Native Americans off cornmeal, and onto white wheat flour and baking powder breads."[23]

In all three of these cases—sugar and tea in the poorhouses of Britain, pickles in America's immigrant populations, and corn in reservation rations—the real threat seemed to be that the poor might enjoy their food. Culinary pleasure was luxury and potential moral hazard that only the middle and upper classes were entitled to. The historical debasement of bodily pleasures that prevented food from being taken seriously in philosophy that Valgenti discusses may have played a role, too.

Instead of the universal nutritional prescriptions we're more accustomed to, cooking instruction and nutritional advice at the turn of the twentieth century was often explicitly tailored to different income levels and occupational types. Ellen Richards, one of the pioneers of the American domestic science movement, said that food for professors, students, doctors, and teachers should be "liberal, varied, well-cooked, and, especially, well-flavored" and "delicately served with all the attractiveness of napery and china," while manual workers should be fed hearty food in few courses, omitting delicate foods like soup or salad. These distinctions were to be maintained even when nonmanual laborers engaged in strenuous physical activity. One of the quotations from Richards that Charlotte Biltekoff also mentions is illustrative: "While the soldier could be given a ration of bread, bacon, and beans . . . the Harvard boat crew required some 'frills,' such as ice cream and strawberry shortcake."[24] Accordingly, home economics lessons aimed at different demographics taught different skills and recipes: "While public school girls learned about baked beans and Indian pudding, middle-class ladies learned from Fannie Farmer and others not just the principles of scientific cooking but the pleasures of preparing and serving all-white or all-pink meals."[25] While the working-class girls were schooled in stretching cuts of meat in stews, middle-class girls learned how to create dainty composed salads and blanket meals in white sauce, which was seen as a "civilizing influence" that could purify and ennoble everything from humble vegetables to Frankfurt sausages.[26]

When they weren't criticizing poor people's diets or using menus to affirm class distinction, home economists looked to the poor for inspiration in developing cost-efficient and palatable meals and praised their frugality and resourcefulness. In the same book where she criticized the Eastern European penchant for pickles, Bertha Wood wrote somewhat more complimentary things about Mexican food, like, "When not too highly seasoned, Mexican dishes are very tasty. . . . Only lack of variety and the use of hot flavors keep their food from being superior to that of most Americans.[27] During the World Wars, nutritionists often looked to immigrants and the poor for ideas about how to stretch cheaper cuts of meat or prepare meals without it. The Common Council for American Unity, a group founded in 1940 with the mission of increasing popular appreciation for the contributions of America's ethnic minorities, noted that "foreign-born housewives are more careful both in their food-buying habits and in the economical preparation of food" and published a brochure called "War-Time Recipes Used by Our Foreign Origin Americans," including recipes like Czech lentil chowder, Italian *pasta e fagioli*, Norwegian smørrebrød, and Polish *golarki* and *pierozki*.[28]

The 1960s counterculture, often seen as a forerunner of today's food movement, also idealized ethnic foods and the food of the poor. The hippies who embraced "natural" foods often turned to ethnic markets for cheap bulk commodities like dried beans and whole grains. They also romanticized "Old World" traditions and "peasant cuisines" that relied on little to no meat, and especially little of the prime beef so central to mainstream midcentury American cuisine. According to Warren Belasco, ethnic foods also "seemed closer to the earth, nature, life, and death."[29] Any affiliation (real or imagined) with the global poor was also attractive to activists opposed to U.S. imperialism and capitalism. It was in this social milieu that Alice Waters first found a receptive audience for the French home cooking traditions emphasizing fresh, in-season ingredients that she helped popularize at Chez Panisse in Berkeley, California. By the 1980s, the menus of trendy and expensive restaurants catering to urban professionals on both coasts were littered with items described as "peasant" fare. As Rachel Laudan argues, it became a mark of sophistication to resist "modern" food. This romantic fetishization of the poor continues in the food movement today; I recently received an email promoting a new series of cookbooks put together by the Rome Sustainable Food Project which advertised their devotion to *la cucina povera*, "the food of the poor, a style of simple, wholesome, and wonderfully flavorful cooking that developed centuries ago in the agricultural communities of Italy."[30] They would likely struggle to sell a cookbook based on the actual diet of the Italian poor in centuries past, who ate monotonous meals of grain-based porridge and root vegetables, just as Laudan argues most peasants have the world over for most of human history.

What explains this diversity of attitudes toward the foods of the poor? It's probably related to the diversity of attitudes toward the poor themselves. Sometimes the elite finds it expedient to characterize the poor as profligate and intemperate, personally responsible for their own penury and any deficiencies in their diet. At other times, the poor become a foil for the elite's anxieties about their own wealth and excesses, representing a virtuous ideal from which the elite has fallen. Romanticizing and elevating foods specifically associated with the poor may also help mitigate anxieties about food snobbery. The "peasant" dishes on an expensive restaurant's menu may serve to demonstrate that the gourmet aesthetic isn't just about scarcity and status, but a kind of culinary meritocracy: any food, no matter how humble its origins, can be considered gourmet if it tastes good enough.

The history of the term "snob" also dates back to that crucial period of class re-formation between the mid-eighteenth and mid-nineteenth centuries. Around the same time Hume and Kant were publishing their theories on universal taste, "snob" entered colloquial use in England, initially to refer to the "ordinary" classes or people without titles.[31] By the mid-nineteenth century, when William Thackeray published a collection of satirical profiles called *The Book of Snobs*, the term had narrowed to refer specifically to ordinary people who imitate nobles. Emulation of the rich was nothing new. Sumptuary laws designed to prevent commoners from dressing like aristocrats and make sure disfavored groups like prostitutes were clearly identified date at least to the seventh century B.C.E. in Greece. By the thirteenth century, English law specified what colors and types of fabric and trim were allowed to persons of various ranks or incomes.[32] The Colonial Laws of Massachusetts Bay passed in 1651 stipulated that anyone with personal fortune of less than £200 who dared to wear gold or silver lace, gold or silver buttons, or silk hoods or scarves could be fined ten shillings.[33] However, these laws were often poorly enforced and openly flouted, especially after the Industrial Revolution made it possible for a much larger population to consume goods and wear clothes that had previously been seen as the exclusive privilege of the aristocracy. As the dismantling of the feudal class system and rise of the new bourgeoisie continued, "snobbery" began to acquire a second meaning: someone who looks down on people of a lower social status. The term also began to lose its initial association with commoners, or people without titles. The result is that today, the word "snob" refers both to ordinary people who commit the sin of reaching above their station and to anyone—rich or not—who sneers at people they deem inferior.

By acknowledging that class status is at least in part about *performance* rather than some sort of *essence*, snobbery undermines status hierarchies. The very idea of snobbery exposes the instability and socially constructed nature of class distinctions. It reminds people that, theoretically at least, anyone could pass as richer or more prestigious than they are, so the signs of class status might be misleading. At the same time, the act that "snobbery" refers to reinforces class hierarchies by investing in the social value of the cultural markers (like gold lace and buttons) that communicate class status. It suggests that it really *is* better to have and do the things the elite has and does, and class hierarchies are not just arbitrary social constructions. Furthermore, the word's pejorative nature attests to the risks of trying to pass and failing. There might be rewards to performing elite cultural capital if you succeed, but falling short and being outed as a "snob" means you're a fraud who doesn't really belong. The word conjures up the possibility of class transgression and punishes the failure to pass in one syllable.

The popular adoption of the word "gourmet" in English in the early nineteenth century is also suggestive of a growing concern about the relationship between food and social class at that time. The first citation for "gourmet" in the *Oxford English Dictionary* is Ange Denis Macquin's 1820 "*Tabella Cibaria*, The Bill of Fare: A Latin Poem, Implicitly Translated and Fully Explained in Copious and Interesting Notes."[34] In the notes explaining the poem, Maquin distinguishes between the "gourmet," who has a "merely theoretical" interest in food; the "glutton," who simply loves to eat; and the "gourmand," who "unites theory with practice." He claims that the French term *gourmand* is equivalent to "*epicure* in the full sense of the word, as we use it in English," but says no existing term in English corresponds to the French *gourmet*.[35] So from the beginning, "gourmet" carried a connotation of affectation: someone who pretends to appreciate food because of what it represents but has no real appreciation for the gustatory and bodily pleasures of the table.

This distinction between *gourmets* and people who actually like to eat prevailed in the early use of the term. In 1835, Washington Irving described people in the American West eating "with an appetite unknown to the gourmets of the cities."[36] An 1851 article titled "Gastronomy and Civilization" in *Fraser's Magazine* specifically associates the term with the elite: "Paul de Kock represents an age when the pretension to gastronomical enjoyment is as universal as liberty, equality, and fraternity; from the discriminating gourmetise of the young nobleman, to the expansive gourmandise of the voracious grisette, all more or less gastrological."[37] Although united by their "pretension to gastronomical enjoyment," "gourmand" and "gourmet" are positioned here on separate ends of the gastrological continuum, represented on one end by the hungry working-class girl who presumably enjoys food for its taste and filling qualities and on the other by the fastidious nobleman more concerned with appearances.[38]

What seems notable about the emergence of both the terms "snob" and "gourmet" around the turn of the nineteenth century—at the same moment Biltekoff argues the American middle class was coalescing and seeking ways to distinguish itself—is that the phenomenon of using cultural cues to perform a desirable class status must have been common enough to require language to describe it. Additionally, the fact that both terms were used as insults suggests this symbolic class-climbing was looked down on. The terms don't just name the behavior, they also police it. Anxieties about snobbery and the social policing of class-climbing complicate the appeal of foods associated with high status. There may be rewards for learning to eat, like, and talk about high-status foods as long as it's interpreted as genuine, but if it comes off as pretentious, there may be negative social consequences.

A cookbook called *Dining for Moderns with Menus and Recipes: The Why and When of Wining* published in 1940 by the New York Women's Exchange exemplifies these dual pressures. The book is aimed at the aspirational middle classes, saying that women who all do their own housework and just want to know how to let "food cook itself" or those rich enough to have a butler and kitchen staff should return the book for a refund.[39] Instead, it's designed for a household with one or two full-time servants and a hostess who "frequently entertains—either through choice or force of circumstances—and knows how her dinner should turn out. But she doesn't quite know how to go about it." The menus are all followed by notes about the recommended wine pairings with advice like, "Good wine should be served with this dinner" or "A bottle of well matured Burgundy, a Pommard or a Gevery-Chambertin of such a good year as 1929 will be just the thing."[40] An introductory message from the wine expert who contributed to the book recommends splurging on rich wines when possible, "even to a Sauternes with the dessert." However, rich wines must be deployed judiciously: "Please don't get your guests vinously bored by serving Sauternes throughout a meal."[41] Additionally, he warns against serving wine "simply to show off," saying it would be better to "buy all the most expensive sparkling wines you can get and die early of indigestion." The book explicitly seeks to help women impress their guests with food and wine, yet insists they must not use wine just "to show off." The admonishment exemplifies how fraught the process of performing elite cultural capital can be.

What the preceding examples suggest is that the contemporary debate about elitism in the food movement is just one more iteration of longstanding anxieties caused by the class implications of *good* food. What does that mean for us now? Well, for one, the fact that people's ideas about *good* food have probably always been influenced by concerns about status ought to make us more critical and self-reflexive about how we categorize and rank foods. Are the current elite's preferences for organic produce and artisanal products and disdain for conventional, mass-produced foods any more defensible than the home economists' preference for bland New England fare over spicy, pickled immigrant cuisines? Alas, there is no reason to believe they are.

The "food revolution" sometimes has the appearance of a movement unified by a coherent ideology. Shopping at farmers markets seems to go hand in hand with buying free-range meat, avoiding high-fructose corn syrup, drinking craft beer, and knowing how to appreciate exotic "superfoods" like quinoa and açai berries. These "alternative" foods are supposed to be healthier, tastier, more environmentally friendly, and more authentic than standard American fare. However, on closer examination, the ideals of the "food revolution" aren't always so compatible. If buying local is really superior, exotic imports like quinoa and açai (not to mention many "gourmet" cheeses, wines, spices, oils, vinegars, teas, and coffee) ought to be deemed inferior. Similarly, although it might be possible to cook a meal that's "gourmet" and "authentic" and still satisfies most people's concept of "healthy" eating, frequently the former involves the use of "fattening" ingredients like meat and cream and a progression of courses that cumulatively contain far more calories than an average fast food meal.[42] Alternatively, some people might find steamed kale with tofu and brown rice delicious, but that wouldn't fit many people's definitions of "gourmet," nor is it "authentic" to any historical, regional, or ethnic culinary tradition. Unless you happen to live within a hundred miles of both rice paddies and a tofu factory, it's not likely to be "local" either.

What seems to unify the many kinds of food characterized as superior in the current food revolution isn't any coherent ideology, but their distinctiveness from the mainstream. That's a good sign that what's elevating them in some people's minds has more to do with status than rationale or some kind of objectively pleasing taste—and here I depart from Ken Albala's argument that it is "great taste" that primarily explains the rising popularity of artisanal products. Instead, I suspect that most working-class and poor Americans would reject what both he and Laudan call culinary Luddism even if they had the time and money to partake in it. This is not to say that they do not cook; on the contrary, low-income Americans eat meals prepared at home the most frequently, and the poorer they are, the more likely they are to cook those meals "from scratch."[43] However, their scratch cooking is likely born more of the necessity that Laudan argues has always forced poor people to do laborious food preparation than any resistance to what Albala calls "industrial pabulum." Where I agree with Albala is in his insight that one of the rewards of cooking for some people is the possibility of gaining social standing, although that may only be available to the classes who reject industrial food on aesthetic grounds. Indeed, the widespread anxiety about elitism in the food movement may in part be a tacit acknowledgment of how much the movement is driven by people using *good* food to try to perform a desirable class identity. The preoccupation with food snobbery seems to reveal the guilty conscience of the food revolution.

The history of status-conscious consumption also offers some insight into why many people might not be so eager to buy into the new diet orthodoxy, even if doing so wouldn't cost more than eating fast food (although most poor and working-class people don't do that very often either—fast food consumption rises with income, peaking in American households with incomes in the $60,000–$80,000 range).[44] Nutritionists no longer recommend that working-class people avoid soups and salads, but the hierarchies that shape our society still influence what kinds of foods appeal to people. Novel, exotic, fancy foods may be unappealing to the working class and poor in part because they're associated with the elite. Not everyone can reap the rewards of performing high-status cultural capital, and the risks and penalties of being called a snob are probably higher the farther you are from the top. The limited market share of organic produce may be due not to ignorance or problems of cost and access, but instead to a more fundamental and long-standing problem with the cultural significance of *good* food. Furthermore, as long as the social construction of good taste is inflected by class politics, it will be difficult for the movement aimed at helping people eat "better" to avoid the taint of elitism.

NOTES

Portions of this essay first appeared in S. Margot Finn, *Discriminating Taste: How Class Anxiety Created the American Food Revolution* (Newark, N.J.: Rutgers University Press, 2017).

1. For brevity's sake, I've opted to use an italicized *good* throughout the chapter to refer to the myriad ways of constructing food as superior associated with the food revolution. It is not meant as an endorsement of that superiority. There is considerable, ongoing debate about what all of these terms mean and whether they're actually better—nutritionally, aesthetically, environmentally, or ethically. The italics, then, are intended as a reminder that this sense of *goodness* in food is a social construction.

2. Andrew Martin, "Is a Food Revolution Now in Season?," *New York Times*, 21 March 2009, https://www.nytimes.com/2009/03/22/business/22food.html.

3. Catherine Greene, *Growth Patterns in the U.S. Organic Industry*, U.S. Department of Agriculture, Economic Research Service, 24 October 2013, https://www.ers.usda.gov/publications/pub-details/?pubid=42456.

4. "Is the Organic Food Movement Elitist?," *Plenty Magazine*, 1 April 2007, republished on *Mother Nature Network*, 1 June 2009, https://www.mnn.com/your-home/organic-farming-gardening/stories/is-the-organic-food-movement-elitist.

5. Ian Brown, "Author Michael Pollan Explains the War on Food Movement," *Globe and Mail*, 18 March 2011, https://www.theglobeandmail.com/life/author-michael-pollan-explains-the-war-on-food-movement/article573363/.

6. Kim Severson, "Some Good News on Food Prices," *New York Times*, 2 April 2008, https://www.nytimes.com/2008/04/02/dining/02cheap.html.

7. "Is the Organic Food Movement Elitist?"

8. Mark Bittman, "Is Junk Food Really Cheaper?," *New York Times*, 24 September 2011, https://www.nytimes.com/2011/09/25/opinion/sunday/is-junk-food-really-cheaper.html.

9. Eric Schlosser, "Why Being a Foodie Isn't Elitist," *Washington Post*, 29 April 2011, https://www.washingtonpost.com/opinions/why-being-a-foodie-isnt-elitist/2011/04/27/AFeWsnFF_story.html?utm_term=.5c790393d6bb.

10. Simon Romero and Sara Shahriari, "Quinoa's Global Success Creates Quandary at Home," *New York Times*, 19 March 2011, https://www.nytimes.com/2011/03/20/world/americas/20bolivia.html.

11. David Hume, "On the Standard of Taste," in *English Essays, from Sir Philip Sidney to Macaulay*, ed. Charles W. Eliot (New York: P. F. Collier & Son, 1910), 222.

12. Hume, "On the Standard of Taste," 223–28.

13. Carl Wilson, *Let's Talk about Love: A Journey to the End of Taste* (New York: Continuum, 2007), 81.

14. Pierre Bourdieu, *Distinction: A Social Critique of the Judgment of Taste*, trans. Richard Nice (1979; reprint, Cambridge, Mass.: Harvard University Press, 1984), 6.

15. Bourdieu, *Distinction*, 79.

16. Bourdieu, *Distinction*, 176.

17. Bourdieu, *Distinction*, 7.

18. See Warren Belasco, *Meals to Come: A History of the Future of Food* (Berkeley: University of California Press, 2006), 9.

19. Sidney W. Mintz, *Sweetness and Power: The Place of Sugar in Modern History* (New York: Penguin Books, 1985), 172.

20. Donna R. Gabaccia, *We Are What We Eat: Ethnic Food and the Making of Americans* (Cambridge, Mass.: Harvard University Press, 1998), 123.

21. Bertha M. Wood, *Foods of the Foreign Born in Relation to Health* (Boston: Whitcomb and Barrows, 1922), 90.

22. Gabaccia, *We Are What We Eat*, 124.

23. Gabaccia, *We Are What We Eat*, 130.

24. Charlotte Biltekoff, *Eating Right in America: The Cultural Politics of Food and Health* (Durham, N.C.: Duke University Press, 2013), 41–42.

25. Gabaccia, *We Are What We Eat*, 127

26. Laura Shapiro, *Perfection Salad: Women and Cooking at the Turn of the Century* (New York: North Point, 1986), 91.

27. Wood, *Foods of the Foreign Born*, 9.

28. Gabaccia, *We Are What We Eat*, 145.

29. Warren Belasco, *Appetite for Change: How the Counterculture Took on the Food Industry* (1989; reprint, Ithaca, N.Y.: Cornell University Press, 2007), 63.

30. Angela Hederman, "Seasonal Vegetable Recipes from the American Academy in Rome," email newsletter from the *New York Review of Books*, 21 March 2015.

31. It may also have started off as slang among Oxford students to refer to the townsmen, as opposed to the academic gownsmen. For example, a collection of Cambridge student verse includes a description of a local bookseller from 1781: "Snobs call him Nicholson! Plebian name, / Which ne'er would hand a Snobite down to fame / But to posterity he'll go,—perhaps / Since Granta's classic sons have dubbed him Maps." Author unknown, *In Cap and Gown: Three Centuries of Cambridge Wit*, ed. Charles Whibley (London: Kegan Paul, Trench & Co., 1889), 87.

32. Aileen Ribeiro, *Dress and Morality* (London: Berg, 2003), 12–16.

33. Linda M. Scott, *Fresh Lipstick: Redressing Fashion and Feminism* (New York: Palgrave Macmillan, 2004), 24.

34. S.v. "Gourmet," *OED Online*, March 2011, Oxford University Press, http://www.oed.com/, accessed 9 April 2011.

35. Ange Denis Macquin, *"Tabella Cibaria," The Bill of Fare: A Latin Poem, Implicitly Translated and Fully Explained in Copious and Interesting Notes* (London: Sherwood, Neely, and Jones, Paternoster Row; J. Robins and Co. Ivy Lane, Paternoster Row, 1820).

36. The reference is from *A Tour on the Prairies*, the travelogue about his tour of the territories of the West published in 1835. S.v. "Gourmet," *OED Online*.

37. Thomas Carlyle, "Gastronomy and Civilization," *Fraser's Magazine for Town and Country* 44 (December 1851): 605.

38. In keeping with the original sense of the word, "gourmet" was exclusively used as a noun until the early twentieth century. The first citation for "gourmet" used as an adjective is from the Westminster Gazette in 1904: "The public in the matter of jokes is gourmand rather than gourmet." S.v. "Gourmet," *OED Online*.

39. Mrs. G. Edgar Hackey, *Dining for Moderns with Menus and Recipes: The Why and When of Wining* (New York: New York Women's Exchange, 1940), 4.

40. Hackey, *Dining for Moderns with Menus and Recipes*, 13–14.

41. Hackey, *Dining for Moderns with Menus and Recipes*, 6.

42. See, for example, *New York Magazine*'s comparison of the caloric content of the nine-course tasting menu at Per Se and McDonald's Big Mac, which concludes that the Per Se meal alone was approximately equivalent to two and a half burgers (1,230.8 Kcal). Including the amuse bouche, dinner roll, wine, and complimentary chocolates brought the total up to 2,416.2, or four and a half Big Macs. Charles Stuart Platkin, "Per Se, per Calorie," *New York Magazine*, 19 June 2007, http://nymag.com/restaurants/features/31268/.

43. Share Our Strength and APCO Insight, *It's Dinnertime: A Report on Low-Income Families' Efforts to Plan, Shop for and Cook Healthy Meals*, January 2012, https://livewellcolorado.org/wp-content/uploads/2015/09/bg_4_its-dinnertime.pdf.

44. Kim DaeHwan and J. Paul Leigh, "Are Meals at Full-Service and Fast-Food Restaurants 'Normal' or 'Inferior'?," *Population Health Management* 14 (December 2011): 307–15.

5

The Standard of Taste Debate

How Do We Decide What Tastes Best?

CHARLES C. LUDINGTON

"There's no accounting for taste." At least that is the commonly used expression and widely held belief. Because taste cannot be accounted for, the implication of this idiom is that taste is subjective and thus neither right nor wrong, and no particular taste is better or worse than any other. Indeed, few people in twenty-first-century America would publicly argue that an objective standard of taste can be established, whether for food, drink, cars, books, movies, music, romantic partners, or anything else. The idea seems absurd on the face of it. We all have different tastes, but they are just that, different: subjective beliefs that cannot be proved to be true according to any objective criteria. They are not universal but personal. Consequently, there is no to way to rank these preferences according to an objective standard, that is, a standard that exists separate from the person who is judging.

And yet simultaneously we hold a completely contradictory belief. To wit, we unashamedly make pronouncements about our own sensory preferences, and we say that some people have marvelous taste, others have good, indifferent, or poor taste, and still others have horrible taste. We may not always include ourselves in the most flattering category, and we rarely declare our own philistinism unless there is social cachet to be gained from doing so—in which case it's hardly modesty that drives us. But it's clear that we make judgments based on our taste, and about other people's taste, all the time. We do so precisely because we believe taste can be ranked according to qualities that are in the object being judged, not in ourselves. So why then do we still say "There's no accounting for taste" or, if we're feeling learned, *De gustibus non est disputandem* (There is no debating matters of taste)? Do we or do we not believe that taste is subjective and without explanation? Have we been conditioned by our cultural values to express an egalitarian belief even though we hold a hierarchical one? Or is this yet another example of cognitive dissonance: otherwise sane people holding two completely contradictory beliefs at the same time and being blissfully unaware?

The so-called standard of taste debate between these two opposing views is significant even if most people are unaware that they partake in it. For what's at stake is whether the judgments we make about taste have any meaning whatsoever. If taste is entirely subjective, then to state that some object we have just experienced is excellent, wretched, beautiful, or something else communicates nothing other than that we approve or disapprove of it. It says nothing about the object itself. As the Prussian philosopher Immanuel Kant asserted, if everyone has his or her own taste, this "would be equivalent to saying that there is no such thing as taste, i.e. no aesthetic judgment capable of making a rightful claim upon the assent of all men."[1] And if taste is purely subjective, the implications for goodness, truth, and justice—other non-empirically-provable categories—are disastrous. I like this, you like that, who's to say who or what is best? For most things, no danger arises from people having different tastes. The danger only occurs when we think about the broader implications of complete subjectivity, because implicit in the idea of taste are broader cultural beliefs, one of which, of course, is whether there are any ultimate values at all.

In this chapter, I will look at what history may be able to tell us about how taste preferences for food and drink (and by implication other commodities) arise within consumer societies, by which I mean societies wherein the buying and selling of goods is the paramount social and economic activity, and—what usually follows—wherein there is more than one choice of a specific category of food or drink. I will begin by examining the taste for wine in one of the world's first consumer societies, Great Britain, and then investigate the taste for beer in the society that has pushed consumerism to its furthest limits to date, the United States. Both examples illuminate broad if not even universal patterns of how individuals in consumer societies make judgments of taste.

My argument is that while some food and drink, like some literature, art, and music, is discernibly superior to others based on objective criteria that can be articulated, most of what we prefer when we eat and drink is based on our habits and our awareness of the multiple meanings of the objects we consume. So, yes, the objective qualities of the food or drink matter, and so too does our physical and emotional composition at the time of tasting. Taste of and for an object is not entirely a cultural construction, a point with which Margot Finn and Bob Valgenti disagree. But what matters even more is customary practice, how we perceive ourselves in relation to the rest of society, how we want others to perceive us, and what we perceive to be the meaning of the objects we eat and drink. In other words, the subjectivity of taste derives less from our physiological selves than from our social, cultural, and political selves. To be clear, I agree entirely with Margot Finn's assertion that social class and aspirations are the primary determinants of individual taste, as well as her summary of what taste communicates about social status. Indeed, this chapter elaborates on her argument and adds to it by showing the complexity of class influence on taste.[2] However, it also shows that class is not the only factor that influences taste. What follows, therefore, is not a strict chronological history of the taste for wine in England and Scotland, and for beer in the United States. Nor do I assert at what point a certain set of determinants overrides another. Instead, what follows is a thematic history based on what I believe to be universal principles in consumer societies, or, to borrow from the philosopher David Hume, principles that are nearly, if not entirely, the same for all people as they make judgments of taste.[3]

PEOPLE TEND TO PREFER THE TASTE OF WHAT THEY'RE ACCUSTOMED TO

In late seventeenth-century England and Scotland, people who drank wine—and they were mostly men with at least a small amount of disposable income—overwhelmingly preferred red wine from Bordeaux (what they called "claret"). Claret was the least-taxed wine, and thus the cheapest, and tavern goers as well as people with cellars of their own were accustomed to the taste. Of course, being accustomed to something begs the question of how that thing became popular in the first place, and in the case of red Bordeaux wine the answer is that when Eleanor of Aquitaine married King Henry II of England in 1152 C.E., her dowry included both the title of duke and the land of Aquitaine. The vineyards around Bordeaux were therefore "English," even if they were not in England proper. And because tens of thousands of gallons of Bordeaux wine were shipped to England each year, and the Scottish traded with their southern neighbors for wine, both nations developed a taste for red wine from Bordeaux more than any other wine.[4]

After the English lost Aquitaine in 1453 trading links between Bordeaux, England, and Scotland remained in place; French wine was still fiscally favored in both kingdoms, therefore claret remained the most popular wine among all wine drinkers. Over the next two hundred years, the primacy of claret was occasionally challenged by sack (i.e., sherry) from southern Spain, but claret remained at or near the top of English wine imports, and always so in Scotland. However, the situation became complicated in the late seventeenth century, when England and France attacked each other economically and militarily. French embargoes on English cloth were countered by English embargoes on French wine from 1678–85 and 1688–97, which should have meant the end of claret, at least in England. But taste is a stubborn thing. In fact, demand for claret was so strong that many English merchants circumvented the embargoes by smuggling claret, or by declaring it to be Portuguese or Spanish wine.[5] As wine arrived in barrels rather than bottles, and bills of lading could be forged, wine fraud was an easy game and English consumers continued to get the wine they wanted, even if they drank it with a wink and a nudge.

During the second embargo English merchants who specialized in the Spanish and Portuguese trade saw a business opportunity and began to import larger amounts of wine. But still, the overwhelming majority of consumers either drank illegally imported claret or complained when they couldn't find any. The English legal apprentice and hack poet Richard Ames wrote a series of poems in the early 1690s in which the central joke was his futile search for claret and his dislike of the Portuguese wine that was being offered in its stead: "Mark how it smells, methinks a real pain / Is by its odour thrown upon my brain / . . . Fetch us a pint of any sort / Navarre, Galicia, anything but Port."[6] Like Ames, the vast majority of English wine consumers preferred the taste they were used to. That said, the difficulty of getting what you're used to can lead to changes in taste, and this is what began to happen with the taste for wine in England around 1700.

TASTE IS OFTEN INFLUENCED BY POLITICAL VIEWS AND OTHER "TRIBAL" IDENTITIES

English consumers wanted their claret, but by 1697 when the second embargo ended, the English Parliament had increased tariff rates and reversed the old fiscal situation whereby French wines were favored. Instead, French wine now paid £51 per tun (i.e., 252 gallons), Rhenish (German) wine £25, Spanish and Italian wine £22, and Portuguese wine £21. Smuggling and fraud of French wine continued, but anti-French politicians, known as Whigs, did everything they could to push the trade in the direction of the Iberian Peninsula. However, when England went to war with both France and Castilian Spain in 1702, this meant new embargoes, and soon Portugal became England's best, or at least friendliest, wine supplier. This Anglo-Portuguese relationship was cemented in the so-called Methuen Treaty of 1703, which required that the English import tariff on Portuguese wine be at least one-third lower than the import tariff on French wine.[7]

Thus, by the end of the War of the Spanish Succession in 1713, English wine consumers were split between those who still preferred the taste of claret and wanted it back on the market at the lowest possible price, and those who had grown accustomed to the northern Portuguese red wine known as port. As an author for the anti-French journal *British Merchant* wrote in 1713, the war and embargo lasted so "exceeding long, that the Portugal Merchants soon enlarged their trade, and filled the whole Nation with their Wines." But transforming taste was a slow process, as the Portuguese wines "being heavy and strong, did not at first please, and we hanker'd after the old claret of Bourdeaux [*sic*]; but in time the quantities [of claret] wore off, and the Merchants found Ways and Means either to bring the Portuguese Wine to our Palates, or Custom brought our Palates to the Wine: So that we began to forget the French wines, and like the others well enough."[8]

Because most English consumers could neither find nor afford Bordeaux wine during the war years, they were less and less accustomed to it. Meanwhile, they were becoming used to another kind of wine, port, which was made with riper and therefore sweeter grapes and was already beginning to be fortified with a dash of brandy. Just as important, however, many English consumers were happy to get used to port because they could not stomach the idea of buying anything from France, no matter how much they once preferred French wine. As Ames wrote in one of his many poems: "The FRENCH—Altho' indeed no Terrour lye / in the Word *French*, yet there's a strange / and almost unaccountable Antipathy / Against 'em does in English Bosoms range."[9] For many English wine drinkers, the once-familiar taste of claret now contained the bitter taste of hatred.

If English politics were shifting popular taste for wine away from claret at the turn of the eighteenth century, Scottish politics were having the opposite effect; they were strengthening taste for claret. Despite sharing the same monarch with England since 1603 and going to war with England against France in 1689, the independent Scottish Parliament did not place an embargo on French wine until 1701, and only then because the French continued to maintain their embargo on Scottish goods. This embargo, while hardly respected by Scottish merchants—who simply declared their claret to be Spanish or Portuguese wine—pleased both King William and the English Parliament; however, the goal of the Scottish embargo on French wine was to get the French to repeal their embargo. When it was clear that this action would not change French policy, the Scottish Parliament passed the Wine Act of 1703, which allowed for indirect importation of French wines, and this even though Scotland, along with England, was at war with France. The point of the Wine Act was to raise revenue for the government and to send the English a clear message that Scottish foreign policy would not be dictated by London. The English Parliament was furious.[10]

Nevertheless, the English Parliament continued to push for a union with Scotland, which finally occurred in 1707. One of the key elements of the Treaty of Union was that customs duties on imports would be placed at the English level, meaning a significant increase over previous Scottish rates. Scottish merchants scrambled to purchase as much claret as possible before the new rates kicked in, and they also looked to make a profit by selling their cut-rate wine to their soon-to-be-compatriots, the English. English wine merchants were irate, and once again claret was the issue. Nor did the issue go away after the Scottish merchants had made their tidy profit. Faced with higher taxes to pay the war debt, Scots turned to smuggling and fraud. Strategies of all sorts were used to circumvent the law, and the most telling factor was that almost everyone was in on the game. Customs officials could be paid not to notice when a ship landed its cargo, and everyone knew that "Portuguese" wine arriving via Norway was really claret from Bordeaux.[11]

Why did so many Scots drink go along with the fraud and drink illegal wine? First, the Scottish were used to claret; it had been their favorite wine for hundreds of years. As we've seen, people tend to prefer the taste of what they're used to. Second, illegally imported claret came from England's sworn enemy and denied the British government revenue; drinking it was two strikes at once against the juggernaut of Anglicization and what was perceived to be an overbearing English state. Even Scots who supported the British Union, and they were a majority, did not want to have their Scottish identity taken away. For Scottish wine drinkers, the taste of claret could not be separated from the joy of defiance and a belief in an independent Scottish nation, if not an independent Scottish state.[12]

TASTE IS INFLUENCED BY PRICE AND CLASS AFFILIATION

We have seen that many English wine drinkers slowly learned to accept the taste of port and that, for many consumers, its taste was enhanced by a particular definition of English patriotism that included hatred of the French. We have also seen that embargoes and taxes repeatedly increased the price of French claret. Portuguese wine, by contrast, was fiscally favored, making port attractive to anyone on a limited budget. To be sure, the majority of English laborers in the eighteenth century rarely drank wine. It was too expensive. But for middle-ranking consumers, who constituted the majority of wine drinkers, cost also mattered. A Swiss visitor to England, Césare de Saussure, noted in 1726 that the standard price for a bottle of claret in a London tavern was five shillings per bottle, while red port was merely two shillings.[13] Numerous other records show that port was almost always the least expensive wine available to English consumers.[14] Of course, wine of any sort was relatively expensive, French wine especially so. For example, in the 1720s one shilling could get you a simple tavern meal of roast beef, bread, and a pint of beer, including a small tip for the waiter.[15]

It is no surprise, then, that port was the preferred wine of the English middle ranks. It is what they could afford. Records from London livery companies (guilds) in the early eighteenth century show that red port wine was the overwhelming favorite for their banquets. To give but one example, the Ironmongers' Company celebrated the Lord Mayor's Day festivities in 1718 with 150 bottles of red port, 96 bottles of white port, 36 bottles of canary, and 18 bottles of Rhenish (German) wine.[16] In short, mostly port, no claret. Likewise, lawyers, doctors, shopkeepers, and artisans all drank port, and mostly the red variety. So too did students at Oxford and Cambridge, who had earning potential, but not much cash.[17]

While middle-ranking men made port their own, the aristocracy and wealthy gentry continued to prefer claret, which they could afford despite the high cost. This class difference in taste is clearly illustrated by Colonel Robert Walpole, a rustic country squire, and his son, Sir Robert Walpole, the first prime minister of Great Britain and certainly one of its most fashionable men. Colonel Walpole's wine purchases show that he drank almost nothing but red port. Meanwhile, Sir Robert Walpole, who had become one of the richest men in Britain by 1730, preferred the finest and most expensive clarets, Haut Brion, Lafite, and Margaux, reserving the port for his household staff and his "public tables," wherein he entertained the lawyers, merchants, and tradesmen who formed the core of his political support.[18] Preference among wines was inflected by one's political views, but it was also influenced by class status. Port was the wine for the middling sorts. Claret was for wealthy aristocrats and other fashionable elites, whether Whig or Tory.[19]

Of course, this was not the first time that the taste for wine was influenced by social class. In mid-seventeenth-century England, even before the English and French began their long series of embargoes and tariff hikes, the taste for wine, any wine, was largely class-based. When a violent struggle broke out between the supporters and opponents of King Charles I in 1642, wine was clearly on the side of the king. Just as clearly, beer and ale were on the side of Charles's enemies, the Parliamentarians. That does not mean that Royalists never drank beer or ale, or that Parliamentarians never drank wine. But because there was some truth, and more than enough perceived truth to the class affiliations of these drinks, Royalists claimed wine as a symbol or their cause, while simultaneously asserting that all Parliamentarians were urban beer drinkers or rural ale drinkers. In other words, Parliamentarians were allegedly from the middle and lower ranks and were therefore politically illegitimate in a world that still fundamentally believed in aristocratic government.

TASTE IS INFLUENCED BY CLASS IMAGE AND ASPIRATIONS

While taste is influenced by price and reputation, it is also influenced by one's self-image and social aspirations. Elite members of society frequently seek out food and drink that is difficult to procure and expensive when found, as such food and drink is representative of how the elite see themselves and want others to see them: rare and dear. This trend in taste helps to explain why Whig political leaders, while despising France and condemning the importation of claret on a large scale, nevertheless preferred claret to all other wines. But the claret they preferred was not the ordinary claret that had been so popular in England for centuries and for which many people still clamored. Instead, it was a new type of luxury claret that had first been invented around 1660 at a property called Haut Brion just outside the city of Bordeaux.[20] This wine was made from a circumscribed and superior vineyard area, the vines were carefully pruned to concentrate the flavors in the grapes, and once vinified the wine was carefully tended for many months (and soon years), not just weeks, before shipping. In other words, the production cost of this type of claret was significantly greater than the production cost of traditional claret, and the taste of the wine was discernibly different on the palate. Samuel Pepys described wine from Haut Brion in 1663 as having "a good and most particular taste that I never met with."[21] Other Londoners were equally impressed and willing to pay the higher price for Haut Brion claret.

More importantly, once the tariff wars between France and England began, luxury claret was the only sort that was worth the extra expense to purchase, as the tax was the same on all French wines regardless of quality. This meant that if claret producers in Bordeaux wanted to reach the English market, they had an incentive to make luxury claret, and so they did. Of course, smuggled or fraudulently declared claret was still relatively cheap; however, the war of the Spanish Succession (1702–13) made smuggling and fraud more difficult, which gave luxury claret producers a market opening of sorts. The wartime embargo on French wines meant that there was no legal way to ship any sort of claret to England, so in response, luxury claret producers arranged to have their wine "seized" at sea by English corsairs, whereupon the wine was actually paid for but then brought to England as a prize of war by the purchaser, and then sold at government-sponsored auctions. Thus was Haut Brion joined on the English market during the War of the Spanish Succession by wines named Margaux, Lafite, and Latour. These four wines remain among the most expensive and sought-after wines in the world today.[22] And who in England were the customers for this wine in the early decades of the eighteenth century? Obviously, it was very wealthy and fashionable people like John Hervey, First Earl of Bristol; James Brydges, First Duke of Chandos; Charles Spencer, Third Earl of Sunderland; Sir Robert Walpole, First Earl of Orford;, and Queen Anne herself.[23]

Preferring food and drink that represents one's own class image and aspirations is hardly unique to the wealthiest members of society. Others groups do it too, although they are forced by cost to make a virtue of necessity. We have seen that the cost of port made it attractive to middle-ranking consumers. But the fact that early and mid-eighteenth-century port was often a rustic wine with strong tannins and variable alcohol content was not a strike against it; rather, these qualities made port an "honest," "genuine," and "masculine" wine. And, of course, that it was neither French nor Spanish made it a patriotic wine for middle-ranking English consumers who despised France and its ally Spain. Honest, genuine, masculine, and patriotic: these were all ways in which middle-ranking Englishmen saw themselves, in contrast to the aristocracy, whom they accused of being superfluous, artificial, effeminate, and with no regard for the true interests of the nation.[24]

TASTE IS INFLUENCED BY A DESIRE TO GAIN SOCIAL DISTINCTION

Because food and drink have social meaning, and because our place in the social order is never completely fixed, people use taste to gain status within society, or more often within a subset of society. In 1700 luxury claret was a relatively new product, it tasted discernibly different from traditional claret, and it was rare and expensive. For all of these reasons, luxury claret conferred positive social distinction on those who consumed it. It might seem strange that taste for wine mattered in an aristocratic society where social status and political power came with birth, but in reality England was in the midst of rapid social change. Commercial wealth, especially through colonial trade, was increasing every year. The merchant princes were followed by men who made money from money, all of whom insisted on their place at the political table. As opponents of the conservative aristocracy that shunned them, these new "monied-men" were Whigs. There were in fact aristocrats within the Whig party, but in the 1690s the Whigs were very much a blend of old and new, landed and commercial fortunes. As such, they had a legitimacy problem. For what created political legitimacy if not God's fixed, aristocratic order? The Whig answer was taste.[25] Having good taste confirmed the aesthetic sensibilities and therefore the moral credentials of the elite, regardless of their family background. In this equation, good taste was allegedly not about wealth but about the ability to appreciate, discuss, compare, and judge the qualities of consumer objects, to be what the English called "polite."[26] And what better object to appreciate, discuss, compare, and judge than a rare wine that few people could find and even fewer could afford? In short, having good taste and being "polite" was (and is) a form of social and political power.

TASTE IS INFLUENCED BY CONCEPTIONS OF AUTHENTICITY

Because taste for food and drink is influenced by class status, it also poses social risks. Specifically, one's taste must be seen as authentic lest one be deemed a social and political fraud. We have already seen the example of mid-seventeenth-century Royalists who, in the belief that only aristocrats had the right to rule, mocked Parliamentarians as mere beer and ale drinkers, and therefore as political frauds. A century later, however, the tables were turned on the aristocracy and those who aped them. During the early years of the Seven Years' War (1756–63), when Britain and France were once again at war, middle-ranking British social reformers laid the blame for British military and naval losses on the supposed corruption, effeminacy, and thus "Frenchness" of the British ruling elite, who formed the officer class of the army and navy.[27]

According to this critique, the polite taste of the ruling elite for things like luxury claret was precisely the problem because it was not honest, manly, and British taste; instead it was frivolous, feminine, and French. When this critique of elite taste began to stick, British elite men responded by aligning their taste with that of their middle-ranking critics. Elite men's clothing became plainer, darker, with fewer ribbons and ruffles, and elite architecture and design rejected the rococo in favor of more austere neoclassicism. Likewise, elite taste for wine began to turn from luxury claret to middle-class port.[28] After all, according to the middle-ranking critics who needed to be rebutted, to drink port was to drink a glass of true British masculinity, which was of course not an objective measure of masculinity, but masculinity as the English middle ranks defined it. The issue was taste, but the stakes were huge, because if the elite were not true men in the eyes of those over whom they ruled, they would no longer be politically legitimate.

To be sure, if elite men were going to drink port, they demanded a style of port in their own self-image: high quality, aged, and refined, but strong and manly nonetheless. And sure enough, port producers responded. Greater care and cost was put into port production, just as had happened earlier with luxury claret, and the development of the cylindrical bottle in the third quarter of the eighteenth century meant that port could be aged and mellowed for decades without needing to be blended with older wines. Indeed, the first known production of a "vintage" port, bottle-aged wine from a single vintage, was in 1765.[29] With vintage port, elite British men could drink a symbol of middle-class masculinity without having to drink an actual middle-class wine. Everyone was a John Bull now.

TASTE IS INFLUENCED BY GENDER

As the preceding section reveals, the taste for food and drink is also influenced by gender, that is, by the culturally constructed sexual identity of both the consumer and object being consumed. Because men constituted the overwhelming majority of wine drinkers in Britain, wine had for centuries been considered a masculine drink, and the darker and stronger the wine, the more masculine it was. Luxury claret, darker than traditional claret, was a symbol of polite, elite masculinity in the early and mid-eighteenth century. In contrast, red port represented middle-ranking masculinity, which idealized rhetorical bluntness, sartorial simplicity, and physical aggression.[30] Thus, for critics of elite male behavior, claret could not be a manly wine. "No, Sir, Claret is the liquor for boys; port for men; but he who aspires to be a hero must drink brandy," exclaimed Samuel Johnson to his fellow (male) diners at a London dinner party in 1779.[31] He went on to say that that few men were heroic enough to withstand the power of brandy, suggesting that port was manly enough and that claret was not manly at all.

Likewise, in the nineteenth century, after the Napoleonic Wars ended in 1815 with a British (Irish born, actually) field marshal leading the allied armies to victory at Waterloo, the gender of wines caused yet another change in popular taste. As an overtly masculine wine, and by 1780 the overwhelming favorite of both middle-ranking and elite wine drinkers in England and Scotland alike, port was associated with the exceptional drunkenness and violence of the revolutionary era. But in the wake of so many wars, in which Britain ultimately triumphed, aggressively masculine behavior began to be rejected in favor of more peaceful activities in which men and women could partake together. Middle- and upper-class socializing increasingly included both sexes and married couples were obsessed with the idea and practice of "domesticity."[32] In these new, gender-mixed environments, overtly masculine port would not do, and into this breach stepped sherry. If port was the man, sherry was the woman, but as a fortified wine sherry was still strong, and thus worthy of masculine respect. This reputation made sherry the perfect wine for men and women to drink together. Certainly men still drank their port, but mostly when they were alone with other men, while women drank sherry with men, or when alone with other women. Victoria and Albert, the very model of British domesticity, both preferred sherry to all other wines.[33] By the 1840s, sherry and port consumption in the United Kingdom were roughly equal, and by the 1860s sherry had moved ahead.

The mid-nineteenth century marked the first time in British history that women were significant wine consumers, and sherry was their preferred wine, but not theirs alone. As just mentioned, British men from the middle and upper classes also enjoyed supposedly feminine sherry. Politically, they could afford to. After all, British men were on top of the world financially and culturally, and were administering the largest empire the world had ever known; politically speaking, their masculinity was not in question. A dash of femininity was now a dash of refinement, not a sign of masculine decay. Meanwhile, British middle- and upper-class women found in sherry the perfect expression of their desired self-image: feminine and delicate, but not weak. Sherry, whether pale yellow, tawny, or brown in color, tasted of the new gender order.[34]

As the previous narrative is intended to illustrate, social class is the primary—but hardly the only—determinant of people's taste. Moreover, it's my contention that all of the previously mentioned patterns that help to determine taste can be found in consumer societies throughout the world. However, rather than cite multiple examples from far and wide to prove my point, I want to look at one example with which readers of this book might be familiar: the taste for beer in the United States. For those who are unaware, the past hundred years have seen both continuity and, more recently, dramatic change in the taste for beer in America. Specifically, by the time of World War I (1914–18), the dominant style of beer was lager, a type of beer that had been introduced by German immigrants to America in the nineteenth century. Lager took longer to make than English-style ales that had previously dominated American brewing, but it lasted longer in barrel before spoiling and better withstood the rigors of pasteurization once that process was introduced in the late nineteenth century. By 1915 there were already a handful of giant "shipping" breweries such as Pabst, Blatz, Miller, Coors, and Anheuser-Busch, which sent their lager beer far and wide in refrigerated train cars, but there were also some thirteen hundred smaller breweries that produced beer for local markets. All of this changed with the passing of Prohibition in 1919. Most of the small breweries went out of business, while the giant breweries stayed afloat by making dealcoholized beer, malt syrup for "baking cookies" (or home brewing, obviously), and sodas of various sorts. Thus, when Prohibition was repealed in 1933, the big breweries had a nearly clear field in front of them. Indeed, they continued to increase their domination of the market during World War II when the government rationed barley and made it difficult for small brewers to expand production. Moreover, the rationing of barley and the high cost of hops led to the increased use of fillers such as corn and rice, making American-style lager a uniquely light beer long before the introduction of "light" beer in the 1970s, which used its paucity of calories as a selling point for those who wanted to drink a lot of beer.[35]

Post–World War II consolidation of the brewing industry was overwhelming: between 1945 and 1980, the total number of breweries in the United States dropped from 407 to 101, while the five largest producers saw their market share grow from 19% to 76%.[36] Moreover, almost all American breweries, regardless of size, produced a beer that the British beer critic Michael Jackson described in the following way: "They are pale lager beers vaguely of the pilsner style but lighter in body, notably lacking hop character, and generally bland in palate. They do not taste exactly the same but the differences between them are often of minor consequence."[37] Less generous critics of American beer preferred the terms "lawn mower beer" (it only tastes good after yardwork on a hot summer day) or "piss water" (its most notable characteristic is to make one urinate).

Since 1978 much has changed. In that year, President Jimmy Carter signed a bill legalizing home brewing, and very quickly home brewers became businessmen. Microbreweries and brewpubs multiplied like yeast cells in a fermentation vat, so that by 2016, there were roughly thirty-five hundred breweries in the United States, more than in any other country.[38] Portland, Oregon, alone had more than sixty breweries, making it the brewery capital of the world.[39] In 2017, 12.7% of the beer brewed in America was "craft beer," meaning that it was produced by brewers who make fewer than 2 million barrels per year, and almost none of that beer was American-style lager. In that same year, 17.5% of beer consumed in America was imported, and both craft and imported beers have been on an upward trend for the past forty years.[40] There is no doubt that American taste for beer has changed since 1978.

Yet that's only part of the story. In 2017, 69.8% of the beer brewed in America, and three-quarters of all the beer consumed in America, was brewed by the industry giants, and most of that is American-style lager.[41] Bud Light, Coors Light, Miller Lite, Budweiser, and Michelob Ultra (a light beer) were the five top-selling beers by sales figures in 2015, and the top eleven beers on the list were all brewed by A-B Inbev or MillerCoors.[42] In short, while trends are changing, most beer-drinking Americans are drinking the same style beer, and in some instances the same beer, that the majority of Americans have been drinking for more than a century. Just as with English and Scottish wine drinkers in 1700 who wanted traditional, inexpensive claret, many American beer drinkers prefer the taste of what they're used to.

Continuity and change in taste have many causal explanations, and among them, as we've seen, are political beliefs and "tribal" identities. Just as with English consumers who refused to drink French wine because they saw France as the enemy, some political progressives and labor union members in America refused to drink Coors beer from the mid-1960s until the late 1980s because of the hiring practices and conservative political views of leading members of the Coors family.[43] Likewise, in more recent years, political progressives have a much more pronounced taste for craft beers than political conservatives do. Of the ten states with the most breweries in America in 2013, the top eight voted for Barack Obama in both of his election victories.[44] The states with the lowest per capita number of breweries, excepting New Jersey, are all in the politically conservative Deep South and southern Appalachia.[45] These taste trends were confirmed by a study published in the journal *Psychological Science* in 2015, which showed that conservative consumers in the United States tended to prefer established beer brands, while progressives were more likely to try new brands or else older brands whose political views reflected their own.[46] For example, according to a survey from 2012, politically engaged Democrats preferred the taste of Sierra Nevada Pale Ale to all other beers, followed by the taste of "any microbrew." It's not clear why Sierra Nevada in particular was the preference of so many progressives, but the fact that in 2010 the northern California brewery was named the Environmental Protection Agency's "Green Business of the Year" surely has something to do with it. On the other side of the political aisle, Amstel Light, a reduced-calorie form of Heineken, was the preferred beer of politically engaged Republicans, followed closely by Sam Adams Light and Leinenkugel. Disregarding levels of political engagement, Coors Light, Miller Lite, and Michelob Ultra were the biggest sellers on the Republican side, while Corona, Heineken, and Budweiser were the most popular beers among Democrats. For those who are disconsolate about the political divide in the United States in the early twenty-first century, it might come as good news that Bud Light, America's number-one-selling beer, is preferred by an almost equal number of Republicans and Democrats, most of whom admit to being politically inactive.[47]

And then there is Budweiser, which until 2001 was America's top-selling beer when it was unseated by a lighter version of itself. No beer has tried harder to link its "brand" to an idealized image of America than Budweiser, whose red, white, and blue color scheme lends itself to the task. Calling itself the "Great American Lager," Budweiser's advertisements have long used the word "freedom" and have often featured pictures of highways, hamburgers, cowboys on horseback, the Statue of Liberty, and of course the Stars and Stripes itself. In fact, in recent years Budweiser has dispensed with the innuendo and simply painted its cans with the American flag from Memorial Day to Independence Day. This hyperpatriotism coincides with Budweiser's much-publicized donations to private organizations that help support U.S. military veterans with some of the profits from their sales. "No matter what the package color, shape, or size," said the director of Budweiser in 2013, "America's beer supports America's heroes because from May to July, buying Budweiser benefits military families."[48] Budweiser, the company wants you to believe, is the taste of American patriotism, or at least a certain type of American patriotism.

Love of country is a selling point, and so is price, although there's no direct relationship between price and the popularity of beer in the United States. According to 2013 statistics, the average price for a case (twenty-four 12-ounce bottles or cans) of the four top-selling beers was roughly $20. But Anheuser-Busch's Natural Light, Busch Light, and Busch, numbers five, six, and seven by volume, all cost a little bit over $15 per case, so it is either the difference in taste on the palate or class reputation that makes these beers less popular among those who drink the most beer. Similarly, Keystone Light, made by SABMiller, is the least expensive major American brand, at under $15 per case, yet it is only the ninth most popular beer by volume, coming in just after Corona, an import from Mexico, which costs nearly $30 per case.[49] None of this is terribly surprising, as it is an unfortunate American myth that the poor are the heaviest drinkers. In fact, the U.S. Government's Centers for Disease Control reports that adults in families living below the poverty line are 27% less likely to have had a drink in the last thirty days than adults whose household incomes are at least four times the poverty level.[50] Similarly, 68% of college graduates are active drinkers, while that is true for only 35% of high school dropouts, who are also much more likely to be poor.[51]

Having said that the poor in America drink less alcohol, the very poorest members of society who do drink alcohol, and especially those who have a chemical dependency on alcohol, often make the completely understandable decision of getting the most bang for their buck by drinking malt liquor beer. Malt liquor has a higher alcohol content than regular lager beers, is sold at or below the price of the least expensive beers, and is usually sold in larger containers, ranging from 16-ounce cans to 40-ounce bottles. It is also heavily marketed in the very poorest communities, especially among minorities, and its advertising is explicitly meant to appeal to the least powerful members of society by giving consumers, in name, the power, wealth, and social cachet they do not actually have: Steel Reserve, Magnum (a name for a gun, a double-sized wine bottle, and a large condom), Colt .45 (another gun), Panther, Cobra, Hurricane, Schlitz Malt Liquor (the "Bull"), Old English 800, and most telling of all, Country Club. These names, of course, are all ironic jokes by the producers, which to those in the marketing room must be even funnier because the consumers either never get them or actually believe the empowering labels. Alternatively, if the consumers get the joke, they're confirming their own impotence ("Yep, I'm a joke"). Malt liquor names are thus an endless laugh at the indigent, who are often homeless and unemployed and are far more likely to be alcoholics than those who drink regular beer, wine, or spirits.[52] But from the perspective of indigent alcoholics, the taste for malt liquor makes perfect sense. It doesn't require a large investment like a bottle of liquor, and it provides a quick fix or escape from life's woes by its potency. Meanwhile, if you don't get the joke, it provides a pleasant psychological lift, and if you do get the joke (and I suspect most consumers do), it confirms your low self-esteem. After all, taste is a reflection of one's self-image and aspirations.

It doesn't seem necessary at this point to elaborate upon the fact that craft beers, or imported beers, which are either made in smaller amounts or with more expensive ingredients, or simply come from another country, are thus likely to be rarer, more expensive, or more worldly, and thereby reflect the self-image and social aspirations of the majority of their consumers. However, it does seem worthwhile to point out how much image matters when taste is used to gain social distinction. Here is one example. By any measure, Rolling Rock beer is an American-style lager, light in color and body, and best served very cold so as to disguise the fact that it has little flavor. And yet beginning in the 1980s, before the proliferation of microbreweries in America, Rolling Rock used its small-town, small-brewery status to promote itself as a better beer than Budweiser, Miller, Coors, and other mass-produced brands. The implication was that those who noticed the difference in Rolling Rock were themselves superior. "Here's to subtle differences" became Rolling Rock's advertising slogan. However, the only obvious difference between Rolling Rock and America's biggest brands was that Rolling Rock came in painted green bottles with a mysterious number "33" on them. Otherwise, Rolling Rock was just another American-style lager. But in the days before so many craft beers were available, the advertising worked like a charm. Rolling Rock sales skyrocketed in the 1980s. The company was promptly purchased by Labatt Brewing Company in 1987, then the Belgian brewing giant Interbrew in 1995, which became InBev in 2004. Rolling Rock was sold to Anheuser-Busch in 2006, which decided to shut down the brewery in Latrobe, Pennsylvania, and move production to Newark, New Jersey. Rolling Rock's slogan was promptly changed to "Born Small Town." However, with the availability of thousands of craft beers, Rolling Rock has long since lost its ability to provide its consumers with social distinction, and sales have plummeted.[53] And no wonder: for socially and commercially aware craft beer drinkers, it's just another mass-produced, watery, American-style lager to be assiduously avoided.

Which returns us to the topic of taste and authenticity. It should be clear that the conception of authenticity does not exist for only one social class or one type of beer. Everyone seeks to exhibit authenticity through taste, but what is "authentic" is different for different people. For instance, microbrewery beer drinkers believe that their favorite beers are more authentic and therefore taste better because the breweries that make them are independent. These consumers want a story; they want "personality" in their beer. And they want to believe that what they're drinking is what people drank in the past—a past that was, well, more authentic. As Scott Whitley, a MillerCoors executive, said in a 2016 interview for Bloomberg, "Millennials are the most marketed to generation and they know it. Authenticity and heritage, being genuine, is very important to millennials. You don't force anything on them. You let them come to you."[54]

But, of course, it's not about authenticity, it's about the *perception* of authenticity. Because, for example, those who prefer Bud Light, Coors Light, and other top-selling beers do not believe their beers, or they themselves, are frauds. Rather, they believe that they are "keeping it real" by drinking "real" beer, that is, beer without all the pretensions of "handcrafted" this, and "small-batch" that. Thus, they perceive themselves to be more authentic than those who drink microbrews. As one flustered Anheuser-Busch executive stated in a 2014 article for the *Wall Street Journal*, "If you try to be too young and too hip, you lose your base. They'll say, 'That's not my Budweiser anymore.'"[55] So the quest for authenticity always influences people's taste, but authenticity is whatever people believe and want it to be. It is not a fixed category.

Which brings us to gender. First, it should be said that beer in America is masculine, a fact that is proved in many ways. According to a 2014 Gallup poll, 41% of Americans who drink say they prefer beer to all other alcohol, but among men alone that figure rises to 57%, while wine is the preferred beverage of women, at 47%.[56] The masculinity of beer could have much to do with the cultural connection between watching sports and drinking beer, a predominantly male activity, and it may well be related to the fact that beer, far more than wine or liquor, is associated with the respectable working classes, and of working-class men in particular.[57] In that sense, when men need to show that they are "real" men, not men who have airs about them or men who can't control themselves, they drink beer. When President Barack Obama invited the Cambridge police officer James Crowley and Harvard professor Henry Louis Gates (whom the officer had arrested for "breaking in" to his own house) to the White House to share a drink and reconcile their differences, he did not invite them for wine, and certainly not white wine, with all its feminine connotations in America. Nor did he offer bourbon, an American drink and a masculine drink, yes, but perhaps also irresponsible for a midafternoon tipple. Instead, Obama invited them for beer. Notably, Gates, a high-minded public intellectual, drank a light craft beer from his adopted hometown, the nationally available—thus only slightly elitist—Sam Adams Light. Gates may not have known that as a brand, Sam Adams Light slants slightly Republican. Meanwhile Crowley, a working-class policeman who, understandably, may have been intimidated by the erudition of his drinking companions, chose Blue Moon Belgian White. This choice may have been intended to suggest sophistication, but if so, it gave Crowley away. Blue Moon Belgian White is neither a craft beer nor from Belgium; it is brewed by the megabrewery MillerCoors in Golden, Colorado. Finally, Obama, a gifted politician who clearly understood and spoke the American language of beer, chose America's number-one-selling beer, the politically neutral Bud Light.[58] In short, all three men at this very public gathering were striving in their own way and according to their own needs to signal what kind of men they were and what kind of men they were not in their choice of beer, but there was no doubt that they were men.

Likewise, when Supreme Court nominee Brett Kavanaugh was testifying before the Senate Judiciary Committee (and the American public) in 2018 and defending himself against charges of sexual assault, he took the seemingly odd position of grounding his defense in his fondness for beer. In this instance beer, and a lot of it, was a symbol of Kavanaugh's all-American maleness, and he used it to portray himself, not as the perpetrator of a crime, but as a victim. His accuser and critics, intimated Kavanaugh, were not simply attacking him, they were attacking all normal, beer-loving American men. As a rhetorical strategy, it worked.

But what really reveals the masculinity of beer in America, and the effect this masculinity has on taste, is the way beer is marketed. Whether it's the tradition of using bikini-clad women surrounded by beer, or college-aged "dudes" drinking beer and watching sports (often in the company of attractive young women), middle-aged men grilling meat and drinking beer, or handsome older men drinking beer while surrounded by beautiful younger women, we are told in infinite ways that beer is for men, especially men who really, really like women, especially physically beautiful women. Many women who love beer, and salespeople who want more women to love beer, find this situation frustrating.[59] But it will be difficult to undo decades of cultural affiliations between beer, men, and nymphs. Indeed, while women between the ages of twenty-one and thirty-four are more likely than men their age to try craft beer,[60] microbreweries seem to have no intention of emasculating their product; quite the contrary. With craft beer names such as Nutsack Pistachio Stout, Double D Blonde Ale, Polygamy Porter ("Why settle for just one?"), Golden Shower Imperial Pilsner, Panty Peeler Belgian Style Tripel, Busterhiman Cherry Ale, Donkey Punch American Barleywine, and Pearl Necklace Oyster Stout, there is no doubt that beer in America is masculine and will not be gender neutral anytime soon. There is also no doubt that the gender of beer, like custom or habit, political views, "tribal" affiliation, social class, the desire for social distinction, and authenticity, influences the way beer is perceived, the way it tastes, and who has the taste for it.

CONCLUSION

I began this essay by asserting that objective taste is possible so long as criteria for judgment are clear, but I have argued that most of what determines our taste is subjective. Taste is not primarily subjective because our individual mouths, noses, tongues, and stomachs are unique, although this certainly influences our sense of taste. Instead, taste is subjective because we are conditioned through practice to prefer certain foods and drinks and because we are aware of our own multiple positions in society, as well as the multiple personal and societal meanings of the things we eat and drink. In that sense, it is best to understand the subjectivity of taste as intellectual and emotional rather than physical. Understanding and accepting this sort of subjectivity allows us to see how our own reality has been constructed first and foremost by forces outside our control, although forces to which we can react. And while I maintain that some things really do taste objectively better than others, we should also remember that having good taste is not an indicator of moral superiority, as most taste is culturally conditioned in the ways that I have shown. Having good taste does not make a person good. It follows that taste judgments should not be moral judgments, at least not in most categories of taste.[61] However, as this chapter has also attempted to show, in consumer societies—where taste positions us—separating taste from morality is impossible to do.

NOTES

1. Immanuel Kant, *Critique of Judgment* (1790), trans. James Creed Meredith (Oxford: Oxford University Press, 1928), 52.

2. The theoretical works that have informed this essay are numerous, but most important of all has been Pierre Bourdieu, especially his book *Distinction: A Social Critique of the Judgment of Taste*, trans. Richard Nice (Cambridge, Mass.: Harvard University Press, 1984). Other influential works are Priscilla Parkhurst Ferguson, *Accounting for Taste: The Triumph of French Cuisine* (Chicago: University of Chicago Press, 2004); Steven Mennell, *All Manners of Food: Eating and Taste in England and France from the Middle Ages to the Present* (Champaign-Urbana: University of Illinois Press, 1996); Carolyn Korsmeyer, *Making Sense of Taste: Food and Philosophy* (Ithaca, N.Y.: Cornell University Press, 1999); Antoine Hennion, "Those Things That Hold Us Together: Taste and Sociology," *Cultural Sociology* 1, no. 1 (2007): 97–114. Finally, multiple chapters in Barry C. Smith, *Questions of Taste: The Philosophy of Wine* (Oxford: Oxford University Press, 2007), have influenced my views on the subjectivity/objectivity debate.

3. David Hume, "Of the Standard of Taste," in *Essays: Moral, Political, and Literary* originally published 1758 (London: Henry Frowde], 1904), 246.

4. For Anglo-French trade in the Middle Ages, see Yves Renouard, "Le grand commerce de vins de Gascogne au Moyen Age," *Revue Historique* 221 (1959): 261–304; Susan Rose, *The Wine Trade in Medieval Europe 1000–1500* (London, 2011), 59–88. For the Scottish medieval wine trade see S. G. Edgar Lythe and John Butt, *An Economic History of Scotland, 1100–1939* (Glasgow: Blackie, 1975), 62–69.

5. Charles C. Ludington, *The Politics of Wine in Britain: A New Cultural History* (Basingstoke: Palgrave Macmillan, 2013), 31–45.

6. Richard Ames, *The Bacchanalian Sessions, or the Contention of Liquors with a Farewell to Wine* (London, 1693), 23.

7. For details about the Methuen Treaty (or, more precisely, Treaties), see Alan David Francis, *The Methuens and Portugal, 1691–1708* (Cambridge: Cambridge University Press, 1966), 184–218; Ludington, *Politics of Wine in Britain*, 62–64.

8. Charles King, ed., *The British Merchant, or Commerce Preserv'd*, 3 vols. (London: John Darby, 1721), 2:277.

9. Richard Ames, *The Double Descent* (London, 1692), 24.

10. For a longer discussion of the Wine Act of 1703, see Ludington, *Politics of Wine in Britain*, 50–53.

11. For an extensive examination of wine smuggling in eighteenth-century Scotland, see Ludington, *Politics of Wine in Britain*, 104–13.

12. For the meaning of claret in Scotland, see Ludington, *Politics of Wine in Britain*, 113–18.

13. César de Saussure, *A Foreign View of England in the Reigns of George I and George II*, ed. Madame von Muyden (London: J. Murray, 1902), 99–100.

14. Ludington, *Politics of Wine in Britain*, 129–30 and passim.

15. Liza Picard, *Dr. Johnson's London: Life in London* (London: Weidenfeld and Nicolson, 2000), 293–98.

16. André L. Simon, *Bottlescrew Days: Wine Drinking in England during the Eighteenth Century* (London: Duckworth, 1926), 74–75.

17. For Oxbridge student wine consumption, see for example James Woodforde, *The Diary of a Country Parson*, ed. John Beresford, 5 vols. (Oxford: Oxford University Press, 1981), 1:13–17.

18. Cambridge University Library, CH (H), "Vouchers, 1657–1745"; John Plumb, *Men and Places* (London: Cresset Press, 1953), 147–52.

19. Ludington, *Politics of Wine in Britain*, 121–43.

20. For Haut Brion and the invention of luxury claret see Ludington, *The Politics of Wine in Britain*, 83–87.

21. Samuel Pepys, *The Diary of Samuel Pepys: A New and Complete Transcription*, ed. Robert Latham and William Matthews, 10 vols. (Berkeley: University of California Press, 1970–83), 4:100 (10 April 1663).

22. Chateaux Haut Brion, Margaux, Lafite, and Latour have consistently been among the most expensive and sought-after wines from the Left Bank of Bordeaux since 1700. In 1855, they were the only four wines to be given "First Growth" status, although they were joined in 1973 by Chateau Mouton Rothschild.

23. For more on the market for luxury claret in the eighteenth century, see Ludington, *Politics of Wine in Britain*, 82–103.

24. For middle-ranking attitudes of themselves and the aristocracy, see Gerald Newman, *The Rise of English Nationalism* (New York: St. Martin's Press, 1987); Kathleen Wilson, *The Sense of the People: Politics, Culture, and Imperialism in England, 1715–1785* (Cambridge: Cambridge University Press, 1995); Michele Cohen, *Fashioning Masculinity: National Identity and Language in the Eighteenth Century* (London: Routledge, 1996).

25. Terry Eagleton, *The Ideology of the Aesthetic* (Oxford: Blackwell, 1990), 31–69.

26. For more on the concept of "politeness," see Lawrence Klein, *Culture of Politeness: Moral Discourse and the Cultural Politics of Eighteenth-Century England* (Cambridge: Cambridge University Press, 1994), and numerous other works by the same author.

27. For more on the alleged effeminacy of the English aristocracy, see Philip Carter, *Men and the Emergence of Polite Society, 1660–1800* (Harlow: Pearson, 2001), 124–38; Newman, *English Nationalism*, 68–84; Wilson, *Sense of the People*, 185–205. For the alleged masculinity of port, see Ludington, *Politics of Wine in Britain*, 153–58.

28. For men's clothing see David Kuchta, *The Three-Piece Suit and Modern Masculinity: England 1550–1850* (Berkeley: University of California Press, 2002); for architecture and design see Michael Snodin, "Style in Georgian Britain, 1714–1837," in *Design and Decorative Arts, Britain 1500–1900*, ed. Michael Snodin and John Styles (London: V and A, 2004, 198–203. For wine, see Ludington, *Politics of Wine in Britain*, 144–62.

29. Christie's (Company Archives, St. James, London), catalogue for an auction held 1 April 1773.

30. Ludington, *Politics of Wine in Britain*, 121–62.

31. James Boswell, *The Life of Samuel Johnson*, ed. R. W. Chapman (Oxford: [add publisher], 1980), 1016.

32. John Tosh, *A Man's Place: Masculinity and the Middle-Class Home in Victorian England* (New Haven, Conn.: Yale University Press, 1999); Mary Poovey, *Making a Social Body: British Cultural Formation, 1830–1864* (Chicago: University of Chicago Press, 1995).

33. For Victoria and Albert's taste in wine, see, for example, British Library, Add. MS 76683, Summary of Monthly Returns Made by the Gentleman of Her Majesty's Wine Cellar, 1 April–31 December 1856, or, more comprehensively, Charles C. Ludington, "Drinking for Approval: Wine and the British Court from George III to Victoria and Albert," in *Royal Taste: Food, Power, and Status at the European Courts after 1789*, ed. Danielle de Vooght (Farnham: Ashgate, 2011), 57–86.

34. Ludington, *Politics of Wine in Britain*, 221–37.

35. For a good general history of brewing and beer in America, see Maureen Ogle, *Ambitious Brew: The Story of American Beer* (Orlando: Harcourt, 2006).

36. Beer Institute, *Brewers Almanac* (Washington, D.C.: Beer Institute, 1993), 7–8; Walter Adams and James Brock, eds., *The Structure of American Industry*, 9th ed. (Englewood Cliffs, N.J.: Prentice Hall, 1995), 125.

37. Michael Jackson, *New World Guide to Beer* (Philadelphia: Running Press, 1988), as quoted in "The American Beer Story," http://www.craftbeer.com/the-beverage/history-of-beer/the-american-story, accessed 1 February 2016.

38. Brewers Association, "Number of Breweries," https://www.brewersassociation.org/statistics/number-of-breweries/, accessed 1 February 2016.

39. Oregon Brewers Guild, "Fact Sheet," 1 July 2015, http://oregoncraftbeer.org/facts/.

40. Brewers Association, "National Beer Sales and Production Data," https://www.brewersassociation.org/statistics/national-beer-sales-production-data/, accessed 30 November 2018.

41. Brewers Association, "National Beer Sales and Production Data," https://www.brewersassociation.org/statistics/national-beer-sales-production-data/, accessed 30 November 2018.

42. Joshua Malin, "The 20 Most Popular Beers in America," Vinepair, http://vinepair.com/wine-blog/20-most-popular-beers-america/, accessed 30 November 2018.

43. Dan Baum, *Citizen Coors: An American Dynasty* (New York: Harper Collins, 2000); B. Erin Cole and Allyson Brantley, "The Coors Boycott: When a Beer Can Signaled Your Politics," Colorado Public Radio, 3 October 2014, http://www.cpr.org/news/story/coors-boycott-when-beer-can-signaled-your-politics.

44. Niall McCarthy, "California Is America's Craft Beer Capital," Statista, 19 May 2014, https://www.statista.com/chart/2262/california-is-americas-craft-beer-capital/.

45. "Craft Beer Breweries per Capita in the United States, by State," Statista, http://www.statista.com/statistics/319978/craft-beer-breweries-per-capita-in-the-us-by-state/, accessed 1 February 2016.

46. Romana Khan, Kanishka Misra, and Vishal Singh, "Ideology and Brand Consumption," *Psychological Science* 24, no. 3 (March 2013): 326–33.

47. National Media: Research, Planning, and Placement, http://natmedia.com/2011/01/08/the-politics-of-beer/, accessed 1 February 2016 (page discontinued).

48. Anheuser-Busch, http://anheuser-busch.com/index.php/starting-today-every-budweiser-sold-through-july-4-will-benefit-military-families/, accessed 1 February 1 2016 (page discontinued).

49. Malin, "The 20 Most Popular Beers in America."

50. U.S. Department of Health and Human Services, Centers for Disease Control, *Health Behaviors of Adults: United States, 2005–2007*, Vital and Health Statistics Series 10, no. 245 (2010).

51. Substance Abuse and Mental Health Services Administration, Office of Applied Studies, *National Survey on Drug Use and Health*, vol. 1, *Summary of National Findings* (Rockville, Md.: U.S. Department of Health and Human Services, 2010).

52. Ricky M. Blumenthal, Didra Brown Taylor, Norma Guzman-Becerra, and Paul L. Robinson, "Characteristics of Malt Liquor Beer Drinkers in a Low-Income Racial Minority Community," *Alcoholism: Clinical and Experimental Research* 29, no. 3 (March 2005): 402–9.

53. Jay Livingston, "Upwardly Mobile Beer: The Class Status of Rolling Rock," *Sociological Images* (blog), Society Pages, 28 March 2013, https://thesocietypages.org/socimages/2013/03/28/upwardly-mobile-beer-rolling-rock-and-the-working-class/.

54. As quoted in Jennifer Kaplan, "Big Beer's Plan to Sell to Consumers Who Hate Them," *Bloomberg Business*, 13 January 2016, http://www.bloomberg.com/news/articles/2016-01-13/big-beer-s-curious-plan-to-sell-to-consumers-who-mistrust-them.

55. As quoted in Tripp Mickle, "Bud Crowded Out by Craft Beer Craze," *Wall Street Journal*, 23 November 2014, http://www.wsj.com/articles/budweiser-ditches-the-clydesdales-for-jay-z-1416784086.

56. Lydia Saad, "Beer Is Americans' Adult Beverage of Choice This Year," Gallup, 23 July 2014, http://www.gallup.com/poll/174074/beer-americans-adult-beverage-choice-year.aspx.

57. Lawrence A. Wenner and Steven J. Jackson, eds., *Sport, Beer, and Gender: Promotional Culture and Contemporary Social Life* (New York: Peter Lang, 2009).

58. Jake Tapper, Karen Travers, and Huma Khan, "Obama, Biden Sit Down for Beers with Gates, Crowley," ABC News, 30 July 2009, http://abcnews.go.com/Politics/story?id=8208602&page=1 http://abcnews.go.com/Politics/story?id=8208602, accessed 1 February 2016.

59. Chad Walsh, "Have You Really Come a Long Way, Baby? How Beer Is(n't) Marketed to Women," Beer West, http://www.beerwestmag.com/the-magazine/feature-have-you-really-come-a-long-way-baby/, accessed 1 February 2016.

60. "Share of Weekly Craft Beer Drinkers in the United States in 2015, by Generation," Statista, http://www.statista.com/statistics/289529/us-craft-beer-drinkers-by-age-and-gender/, accessed 1 February 2016.

61. For a rebuttal of this argument, see Robert Valgenti, "Is Thinking Critically about Food Essential to a Good Life?" (chapter 11 in this volume).

6

What Does It Mean to Eat Right?

Nutrition, Science, and Society

CHARLOTTE BILTEKOFF

Perhaps nothing better captures today's politics of nutrition better than what we might call the beverage wars.[1] I am not thinking of the advertising battle between Coke and Pepsi, an expensive duel over consumer preference and loyalty that ultimately floated both boats.[2] The warring beverages I have in mind don't exactly go head to head the way Coke and Pepsi did, and they are not, in the end, both winners. Think instead about two of the most pervasive conversations going on today about what we drink, both of them microcosms for broader conversations about what and how we eat. In one corner, we have profound, and highly politicized angst about the overconsumption of sugary drinks, like Coke and Pepsi, and in the other, joyous indulgence in and celebration of their virtuous opposite: fresh-pressed juices. While economic and cultural elites celebrate home juicing routines, renew body and soul through juice cleanses, and frequent fresh-pressed juiceries (online and in person) selling products such as Mother Earth (cucumber, celery, kale, Swiss chard, dandelion, parsley, lemon, ginger juice: $7 / 11 oz.) and Be Fit (green apple, celery, arugula, fennel, lime, ginger: $10.99 / 17 oz.) the drink of choice among less privileged consumers has come under the scrutiny of intense public health concern, become the target of "sin tax" campaigns (the first of which to be successful recently passed in Berkeley, California, where I live), and been banned from school vending machines.[3] Public health campaigns across the country urge Americans to "rethink your drink" using grotesque images of fat globules being poured out of soda bottles, and dramatic illustrations of the amount of sugar in a large soda (twenty-six packets!) alongside reminders that "all those extra calories can bring on obesity, diabetes and heart disease."

Though simultaneous, these two conversations about beverages take place entirely separately, and they are not usually thought about in relation to each other; public health concern about soda consumption and its effects seems totally unrelated to the growing juicing trend. But, as a historian concerned with broad questions about the cultural politics of nutrition and dietary health, I offer this juxtaposition as a way into thinking about how a historical perspective can help us to see contemporary debates differently. Contemporary concerns about sugary drinks typically raise a fairly predictable—though not uncomplicated—set of questions about the relationship between food (or drinks), health, and social well-being. There are questions about the nutritional composition of sodas, the effect of their consumption on the body, and the relationship between sugary drink consumption and obesity and type 2 diabetes. From a public health perspective, key questions are about the population-level health impacts of the overconsumption of sugary sodas and about how to get people to consume fewer of them. On the level of politics, both the concern about sugary drinks and the various campaigns that have aimed to curtail their consumption and availability raise questions about who is responsible for guarding against poor nutritional choices, the role of the state, the individual, and corporations in ensuring the overall nutritional well-being of individuals and the nation as a whole. Thinking historically connects these beverage politics to a long history of dietary crises and broadens the kinds of questions we must ask.

I propose a framework for thinking through contemporary nutrition politics that is guided by an understanding of the social and cultural role that teaching people to eat right has historically played. Teaching people to eat (or drink) right is never simply about applying the facts of nutrition to improve biomedical well-being. The history of dietary reform in the United States shows, on the contrary, that concerns about bad eating habits typically express broader social concerns and that advice about how to eat right conveys important social and moral ideals. The history of dietary advice in the United States also shows that soda taxes and purifying juices are not unrelated phenomenon. Together, these two related beverage phenomena play a role in ongoing conversations about good and bad food, and good and bad eaters, that have historically contributed to the construction of class in the United States. The morally loaded contrast between good juicers and bad soda drinkers is part of a dynamic between the responsible, health-seeking middle and upper classes and dangerous "unhealthy others"—clearly in need of guidance and reform—that has been part of the American social fabric since the simultaneous emergence of a self-conscious middle class and the modern science of nutrition in the late nineteenth century.

This essay is about the history that can help us make sense of contemporary nutrition politics. Drawing on my work in *Eating Right in America,* I offer three main arguments about the past that challenge and refine the way we think about the politics of eating right in the present. First, I will argue that dietary ideals always serve two functions; they provide empirical rules about how to eat right and ethical guidelines through which people can construct themselves as responsible, moral subjects. This point further illuminates the tensions experienced by the mother's described by Amy Bentley in this volume as they weighed decisions about when to introduce solids, whether to use commercial or homemade foods, and which foods should be introduced first. My analysis suggests that these dietary dilemmas were also moral dilemmas through which mothers might construct themselves as responsible subjects and "good mothers." I will also argue that dietary ideals have historically expressed and conveyed ideals of good citizenship that are class-based. While both dietary advice and cultural notions of what it means to be a good citizen have changed over the course of the last century, the relationship between them has remained consistent. Finally, I will argue that the discourse of dietary reform has played an ongoing role in the construction of class in the United States. Elite ideas about good and bad eaters at the turn of the twentieth century helped to draw a line between the middle and working classes and thus to create distinct class identities. When economic and cultural shifts later in the century led to greater fragmentation of the middle class, eating right played a particularly important role in defining and defending the boundaries of upper end of the middle class, which cast itself increasingly as part of a unified elite. While Margot Finn's argument in this volume usefully points out the role of class distinction and class anxieties in discourses of good food, I argue that ideas about good and bad food not only reflect and create class differences but are also constitutive of the very moral hierarchies on which class itself depends.

In my book, *Eating Right in America,* I use case studies of four dietary reform movements to develop these arguments. These include the turn-of-the-century domestic science movement, a precursor to home economics fueled by Progressive Era social and political dynamics; the National Nutrition Program, a nutrition education program that was part of the wider World War II home front food programs; the contemporary alternative food movement, the origins of which I trace to the 1970s counterculture; and the simultaneous campaign against obesity. In this chapter, I draw on these case studies selectively to explicate the three key ideas described above, that are critical to rethinking contemporary "food fights." My aim is not to provide an exhaustive analysis of contemporary food (or beverage) politics, but to invoke the discourses of green juices and banned sodas as an example of the ways in which the moral categories of good and bad eaters operate in the present. Contemporary food activists and scholars have long embraced the importance of knowing where our food comes from; I aim to show how we can arrive at a more critical understanding of where our dietary advice comes from by thinking through these contemporary figures in relation to the past.

SEEING THE ETHICAL IN THE EMPIRICAL (AND THE EMPIRICAL IN THE ETHICAL)

Let me start by proposing a lens, or framework, through which we might learn to view all instances of dietary discourse. The purpose is to fundamentally change our perception of what nutrition is and does, so that we can engage with the present (and the past) more critically. Rather than seeing nutrition solely as a set of guidelines for choosing a healthy diet, I argue that we should learn to see its dual function. Nutrition, as John Coveney has argued and as my own research has shown, is always both empirical and ethical; it provides rules about what to eat that also function as system through which people construct themselves as certain kinds of subjects.[4] Nutritional ideals reflect social values, so eating right is never simply a biomedical matter. It becomes a means through which people can practice and display highly valued social qualities as well as assess and classify others. Looking at the emergence of the modern science of nutrition and the application of its principles by domestic scientists at the turn of the century elucidates why it's important to keep both the ethical and the empirical in mind when faced with all kinds of dietary discourse, from the USDA's dietary guidelines, to farm to school programs, billboards dramatizing the dangers of sugary drinks, and lifestyle magazine articles about the current juicing craze.

Nutrition emerged as a science in the late nineteenth century through the work of chemists who, by breaking food down to its chemical constituents, provided a quantitative framework for measuring the value of food. Like other sciences establishing their realms of expertise at this time, nutrition staked its claims to authority on the presumed objectivity of the numbers it produced.[5] Those numbers introduced new ways of thinking about food; quantitative frameworks that made it possible to identify a "good diet," measure deviance from ideals, and compare diets to each other. The quantitative scaffolding that the science of nutrition introduced, however, built on existing moral precepts about the management of the appetite. As Coveney shows, before the ascendance of scientific reasoning, the act of choosing what to eat was understood mainly as a matter of ethical or religious concern.[6] Scientific nutrition introduced quantification, but it did not vanquish these moral concerns; rather, it expressed them through its numeric language of dietary ideals.

Wilbur Atwater's table entitled "Comparative Expenses of Food" illustrates this point. Known as "the father of American nutrition," Atwater conducted research throughout the 1880s and 1890s that, building on the work of German chemists, identified the chemical composition of twenty-six hundred common American foods, from very lean and very fat meat to doughnuts and pies, artichokes, lentils, chocolate, parsnips, and figs. He organized the data into tables showing the amount of water, protein, fat and carbohydrate, "ash," and calories or "food value per pound" for each.[7] Later, Atwater developed a quantitative calculation of a "good diet" by putting this information into relationship with two other pieces of data; the cost of food and the energy needs of people engaged in various tasks. Together, he argued, these pieces of information revealed the "true" value of food. Cost, it turned out, was no measure of a food's value. On the contrary, Atwater argued that the best food was that which provided the most energy for work at the least possible cost.

Published in 1888 in the *Century Magazine*, a middle-class monthly, "Comparative Expenses of Food" was part of an article in which Atwater explained these fundamental concepts, and provided the information he believed people needed to choose "good"—that is, economically efficient—diets. He described well-meaning people, like a coal laborer earning seven dollars a week who boasted of giving his family "the best flour, the finest sugar and the very best quality of meat," when cheaper food would have provided his family with just as much "food value." The table provided information designed to remedy precisely this type of "immense economic and hygienic blunder."[8] It showed the amount of nutrition that could be obtained for twenty-five cents from different foods, including beef sirloin, salted codfish, sugar, oatmeal, and rice. The various food materials were listed along the left side of the page, with subsequent columns showing the price per pound, the amount of food that could be obtained for twenty-five cents, and the quantities of protein, fats, and carbohydrates that would be provided by that amount of food, as compared to the daily needs of a man engaged in moderate muscular work.[9]

While the table appears strictly quantitative, a closer look shows that it expressed social values and moral measures in numerical terms. The concept of a good diet that Atwater promoted did not emerge directly from the facts of nutrition, after all. Atwater's laboratory investigations produced information about the nutritive content of foods and the energy needs of humans in general. Atwater put this information together and triangulated it with the cost of food in order to advance a concept of an ideal diet that was directly informed by social concerns. Amid the massive social changes associated with industrialization, urbanization, and immigration, Atwater, like many of his contemporaries among the educated upper-middle classes, was profoundly concerned about the social unrest already under way and the potential for chaos that loomed over rapidly expanding urban centers in the Northeast. His seemingly quantitative dietary ideals served social interests by teaching people to nourish themselves as "efficiently," or cheaply, as possible, thus addressing, for example, worries that laborers would take to the streets to demand higher wages. "Comparative Expenses of Food" expressed the social and moral qualities of thrift, frugality, and responsibility as a quantitative relationship between nutrient value and the cost of food.

The moral dimension of Atwater's calculations may seem like a revelation to us, but it was precisely this aspect of nutrition, and the promise of social reform it contained, that drew turn-of-the-century female reformers to the "Food Problem." Concerned with many of the same social problems that drove their contemporaries to hygiene and temperance crusades, reformers known as domestic scientists believed that improving people's eating habits could improve their morals and elevate their character, and in so doing address many of the most pressing social problems of the time, from intemperance to labor unrest. One reformer writing for a movement magazine in 1894 expressed these intertwined empirical and ethical aims in the following statement: "The relation of nutriment to personal morality is no longer to be ignored. The ministry of diet in the work of character-building is therefore one of the most important studies a woman can undertake."[10] The dual nature of nutrition was clearly at work in domestic scientists' first major reform project, a public kitchen in one of Boston's "poor sections" designed

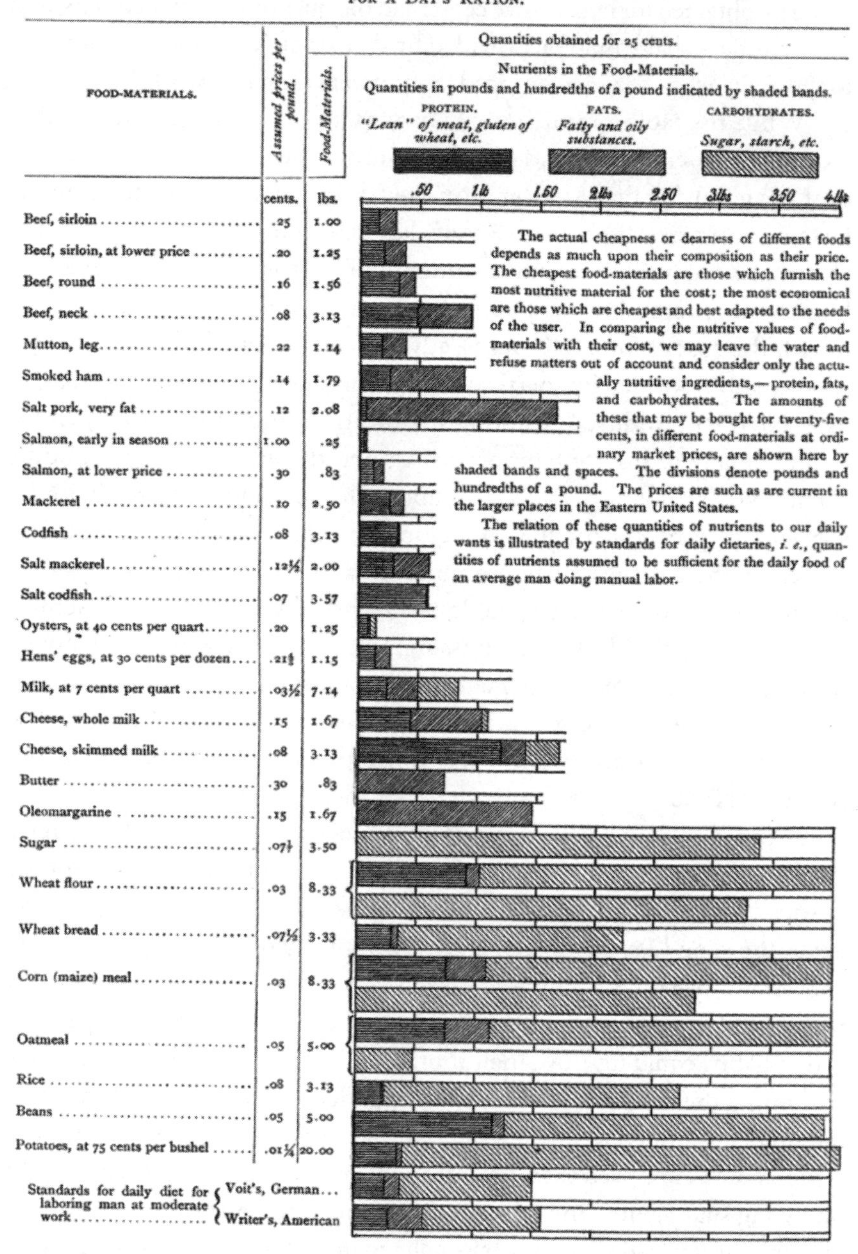

440 DIAGRAM VI.—COMPARATIVE EXPENSIVENESS OF FOODS.

AMOUNTS OF ACTUAL NUTRIENTS (NUTRITIVE INGREDIENTS) OBTAINED FOR TWENTY-FIVE CENTS IN DIFFERENT FOOD-MATERIALS AT ORDINARY PRICES, WITH AMOUNTS APPROPRIATE FOR A DAY'S RATION.

Quantities obtained for 25 cents.

Nutrients in the Food-Materials.
Quantities in pounds and hundredths of a pound indicated by shaded bands.

PROTEIN. "Lean" of meat, gluten of wheat, etc.
FATS. Fatty and oily substances.
CARBOHYDRATES. Sugar, starch, etc.

.50 1 lb 1.50 2 lbs 2.50 3 lbs 3.50 4 lbs

FOOD-MATERIALS.	*Assumed prices per pound.*	*Food-Materials.*
	cents.	lbs.
Beef, sirloin	.25	1.00
Beef, sirloin, at lower price	.20	1.25
Beef, round	.16	1.56
Beef, neck	.08	3.13
Mutton, leg	.22	1.14
Smoked ham	.14	1.79
Salt pork, very fat	.12	2.08
Salmon, early in season	1.00	.25
Salmon, at lower price	.30	.83
Mackerel	.10	2.50
Codfish	.08	3.13
Salt mackerel	.12½	2.00
Salt codfish	.07	3.57
Oysters, at 40 cents per quart	.20	1.25
Hens' eggs, at 30 cents per dozen	.21⅓	1.15
Milk, at 7 cents per quart	.03½	7.14
Cheese, whole milk	.15	1.67
Cheese, skimmed milk	.08	3.13
Butter	.30	.83
Oleomargarine	.15	1.67
Sugar	.07½	3.50
Wheat flour	.03	8.33
Wheat bread	.07½	3.33
Corn (maize) meal	.03	8.33
Oatmeal	.05	5.00
Rice	.08	3.13
Beans	.05	5.00
Potatoes, at 75 cents per bushel	.01⅛	20.00
Standards for daily diet for laboring man at moderate work: Voit's, German		
Standards for daily diet for laboring man at moderate work: Writer's, American		

The actual cheapness or dearness of different foods depends as much upon their composition as their price. The cheapest food-materials are those which furnish the most nutritive material for the cost; the most economical are those which are cheapest and best adapted to the needs of the user. In comparing the nutritive values of food-materials with their cost, we may leave the water and refuse matters out of account and consider only the actually nutritive ingredients,—protein, fats, and carbohydrates. The amounts of these that may be bought for twenty-five cents, in different food-materials at ordinary market prices, are shown here by shaded bands and spaces. The divisions denote pounds and hundredths of a pound. The prices are such as are current in the larger places in the Eastern United States.

The relation of these quantities of nutrients to our daily wants is illustrated by standards for daily dietaries, *i. e.*, quantities of nutrients assumed to be sufficient for the daily food of an average man doing manual labor.

W. O. Atwater's "Comparative Expensiveness of Foods," in which social qualities are expressed as nutritional quantities. (Courtesy of Cornell University Library, Making of America Digital Collection) to address social concerns through scientific nutrition. The New England Kitchen provided carefully calibrated nutritionally "efficient" meals (oatmeal mush, cornmeal mush, beef stew, fish chowder, pea soup, Indian pudding) to be taken home by its patrons. While its offerings were meticulously scientific, its aims were overtly ethical. The reformers saw the open kitchen as a "silent teacher," providing "hints for more healthful living"[11] that would improve people's character and morals. Describing the New England Kitchen in 1891, a prominent domestic scientist celebrated the project's intertwined nutritional and social aims, explaining that when people with poor diets "are educated up to the point where they choose soups, well-cooked cereals, and good milk, there will be a great gain in their physical and moral condition."[12]

Revisiting the emergence of scientific nutrition and modern dietary reform shows that seeing dietary ideal as purely empirical obscures their very important—even intrinsic—moral dimension. It suggests that we should engage with all utterances of dietary discourse as simultaneously ethical and empirical, whether the ethical dimension is obscured by the empirical, as is the case when scientific nutrition is at work, or vice versa (alternative food movements, for example, tend to foreground ethics, but often contain rule-bound empiricism at their core).[13] The difference between a can of Coke and a glass of fresh-pressed kale juice is, on the one hand, empirical; it can be measured nutritionally in the lab. But, on the other hand, the moral dimension of dietary discourse is inevitably at play in how we understand and act on the distinctions between these two beverages. Efforts to reduce soda consumption through taxes and bans are about not only nutrients and biochemistry, but also social and moral concerns. Drinking too much soda is believed to lead to obesity and diabetes, which are medical problems faced by individuals who don't "eat right," that are also perceived as threats to the social and economic well-being of the nation as a whole. Thus, to the rule-setting elites, bad eating habits, like drinking too much Coke, are a quantifiable form of social and moral irresponsibility that indict the lower-middle and working classes and validate the social hierarchy that keeps them on the bottom.

LESSONS IN GOOD CITIZENSHIP

The fundamentally dual nature of nutrition forms a foundation through which dietary ideals express particular social ideals. My research on dietary reform movements has shown that dietary advice has historically conveyed cultural notions of good citizenship. While rules about what constitutes a good diet and ideas about what it means to be a good citizen change over time, the relationship between these two things has remained consistent. During the Progressive Era (ca. 1895–1925), for example, domestic scientists promoted a notion of eating right that helped people to understand and practice the distinct social qualities that were necessary for good citizenship in the context of a growing federal government and increasingly interventionist State.[14] During World War II, nutrition education collapsed ideals of good wartime citizenship onto the figure of the good eater. Looking closely at the dietary lessons of this era provides particularly stark examples of how dietary advice functions as a pedagogy of good citizenship. As these examples will show, the choice between a soda and green juice is a matter of potentially grave social consequence.

Eating right became a national concern during the mobilization for World War II, as new knowledge about the role of vitamins in nutrition and related fears about the impact of the Depression on American's eating habits converged with the imperatives of wartime. The discovery of the vitamins and minerals—invisible, and odorless, yet essential nutrients—led to concern throughout the Depression that even those Americans who seemed properly fed were not eating enough "protective," vitamin-rich foods to fend off the ill effects of malnutrition, leading President Roosevelt to declare one-third of the nation "ill-nourished" in his 1937 inaugural address.[15] By the fall of 1940, with mobilization in full swing and in the wake of alarming Selective Service rejections that many believed to be nutrition related, the question of how to improve American food habits became an official defense issue. The first Recommended Dietary Allowances (RDAs) were launched in 1941 on the eve of the opening of the National Nutrition Conference for Defense, a three-day meeting called by the president to lay the groundwork for a National Nutrition Program that would both promote the new dietary standards and help people to adjust to the exigencies of wartime citizenship.[16]

Throughout the National Nutrition Program, lessons in eating right conflated the qualities of being a good eater with those of being a good wartime citizen. The central guideline of the program was the Basic 7 Food Guide, which on the empirical level translated the highly technical RDAs into simple advice that people could follow in choosing what to eat. But it also clearly operated on an ethical level, presenting eating right as a way for people to align their daily habits with the war effort and practice good citizenship, much like other home front campaigns such as bond drives and scrap collection.[17] The guide was usually presented as a pie chart with seven sections representing groups of nutritionally similar foods from which people were

Empirical and ethical layers of the Basic 7 Food Guide.
(Courtesy of National Archives, photo 44-PA-798B)

supposed to eat every day: green and yellow vegetables; oranges, tomatoes, grapefruit; potatoes and other vegetables and fruit; milk and milk products; meat, poultry, fish, and eggs; bread, flour, and cereals; butter and fortified margarine. A typical representation of the Basic 7 also included a message about the social significance of eating right in the center of the circle; a red, white, and blue image a family of four walking toward the viewer, framed by the motto "U.S. Needs Us Strong: Eat the Basic 7 Every Day." The image of the family, brave and unified, made it clear that the "U.S. Needs Us Strong" referred not only to the imperative for physical strength during wartime but also to the equally important need for social stability and cohesion. Together, the empirical and ethical layers of the Basic 7 Food guide presented eating right as a patriotic wartime duty central to the pursuit of both of these aims.

Throughout the campaign dietary lessons conveyed the idea that good diets were linked to morale, as well as to all of the other qualities and characteristics of good wartime citizenship. In radio briefs read across the New York State airwaves, for example, "poor diets" were described as the possible cause of "lack of courage and willpower," while nutrition was described as essential for reasons that went far beyond the physiological. A radio brief called "Food First," for example, explained, "Courage, cooperativeness, and productive capacity—as well as the good old-fashioned feeling of buoyant and good health—are all very important to national defense. And they all depend upon proper food." In radio briefs, posters, and pamphlets distributed throughout the nation, good eaters were described as having energy, alertness, optimism, resolve, morale, cooperativeness, strength, stamina, clear eyes, good eyesight at night, good posture and "a sense of well being," while bad eaters were depicted as a serious threat to the war effort.[18]

The fate of bad eaters within this construct was dramatized in a pamphlet called *Eat Well to Work Well* that was distributed in defense plants as part of the Nutrition in Industry Program. On one side, the pamphlet showed an image of a strong, cheerful defense worker eating a healthy meal (milk, fruit, and a hearty sandwich) with a headline asking "Worker are you helping Uncle Sam?" and a list of suggestions about how to be a good eater and a good citizen ("Eat a hearty lunch everyday to help keep you in top notch physical condition, to make you feel like doing your job and 'playing ball' with your fellow workers"). On the facing page it showed a scrawny, hollow-cheeked, and vaguely racialized worker lunching on doughnuts and pie next to a headline asking, "Worker, are you helping Hitler?" The juxtaposition made clear that being a bad eater was not simply a biomedical matter. While eating right was likened to being a good wartime citizen, failing to do so was, at least in this case, equated with the very worst form of wartime citizenship: treason. Despite the sarcasm of the "advice" to the workers "Helping Hitler" ("Eat a poor lunch every day so that you will have many accidents, spoil much material, and keep your fellow-workers from getting their work done while you quarrel with them"), this example points to the serious social stakes involved in eating right, or failing to.[19] And despite the tendency to look back on wartime propaganda as a product of a less enlightened, less sophisticated age, there is no question that the relationship between eating right and ideals of good citizenship remains with us today.

Much as the specter of widespread but undetected malnutrition was framed as a threat to the nation as a whole before and during World War II, in more recent years obesity has been represented not merely as a matter of individual health, but as a threat to national well-being that requires the patriotic resolve of "good citizens." The iconography of obesity as a threat to the nation includes the now well-known Centers for Disease Control and Prevention (CDC) maps charting rising rates of obesity in each state since 1985. Moving from cool blues to alarming reds as obesity rose over time, the CDC maps dramatically envisioned the increasing body mass of individual Americans as a danger that was spreading rapidly, like a virus, across the national landmass. Presented in a PowerPoint format and widely disseminated, the maps turned quantitative data about weight gain into an alarming story about a threat to the nation itself.[20]

In the context of post-9/11 social and political realities, obesity discourse took a particularly patriotic turn: now fatness was frequently framed as a threat to both military preparedness and national character. In November 2001, for example, Reuters ran a story titled "U.S. Soldiers Getting Fatter," and early in 2002 a *Washington Times* article explained that overweight troops "can hinder not only their own performance but that of their units as well as the success of their often grueling military missions."[21] A 2005 national weight loss program sponsored by the Discovery Health Network provides striking visual reminders of present-day links between eating right and being a good wartime citizen. In a TV series accompanying the nationwide diet campaign, six contestants "chosen to represent overweight America" were shown dressed in red, white, and blue workout clothes and walking toward the viewer in front of the Capitol Building, much like the stalwart family at the center of the Basic 7 Food Guide. The "body challengers" ran a Marine training course in fatigues for their first challenge; footage of them collapsing in muddy pits was replayed throughout the rest of the series. Challengers wore red, white, and blue gear in their "before and after" shots, and statistics about their weight loss were shown against the backdrop of an American flag.[22]

This history of the relationship between dietary ideals and social ideals is clearly manifest in the contrast between irresponsible soda drinkers and health-seeking juicers, with consequences for the bad eater that are not unlike those for the worker "Helping Hitler." Because of the dual nature of nutrition,

Patriotic iconography in the post-9/11 obesity landscape. (Stills from National Body Challenge, Discovery Health Channel [2004])

conversations about good and bad food, or drinks as the case may be, are always also about good and bad people. Today's "beverage wars" have clear implications, resonating with cultural notions of good citizenship that have attended the late twenty-first-century emergence of neoliberal social and political formations. Soda drinkers, or at least those who drink large amounts of soda, are often represented, for example in public health campaigns, as ignorant and irresponsible—bad citizens whose eating habits threaten the health care system and the economic well-being of the nation as a whole. Juicers, on the other hand, are often represented in elite lifestyle and health magazines as possessing the qualities of self-control, personal responsibility, informed consumption, and health-seeking that are all considered central to contemporary notions of good citizenship.

THE HEALTHY SELF AND THE UNHEALTHY OTHER

The class dynamics involved in the contrast between responsible juicers and dangerous soda drinkers are clearly not far from the surface, and the roots of these dynamics can be traced to end of the nineteenth century. At that time the American middle class was coalescing as a self-conscious social group, but as one historian has explained, even as the term "middle class" was achieving a relatively stable meaning, the concept "crumbled at the touch."[23] Rising wages in some industries and an increase in low-skill clerical jobs blurred class distinctions, leaving the "brain-working" members of the middle class in search of other ways to know and distinguish themselves.[24] As Robert Crawford has shown, the pursuit of health soon became central to a new, more elite middle-class identity, and to its claims of responsibility and authority. The work of creating a distinct identity for what we might call the upper-middle class took place through the production of a dynamic between the "healthy" middle class (i.e., upper and upper-middle classes) and the "unhealthy other" (i.e., lower-middle and working classes) that remains with us today.[25]

As Crawford shows, the identity of the elites as "healthy" was constructed in opposition not just to the concept of disease and death, but also to the imagined "qualities of 'unhealthy' persons already positioned as outside, stigmatized and dangerous."[26] This dynamic can be seen, for example, in the reflections of domestic scientists as they celebrated the "resounding success" of the Rumford Kitchen, a replica of the New England Kitchen that catered to middle-class audiences at the 1893 World's Fair. While the actual New England Kitchen and other public kitchens built on its model were foundering due to lack of patronage in poor and immigrant neighborhoods throughout the Northeast, members of the "intelligent class" flocked to the display Rumford Kitchen to enjoy brown bread and pea soup while reading menu cards showing the relationship between daily nutritional requirements and the composition of each dish being served.[27] Reflecting on the difficulty of changing the diets of "working men," and the ease with which educated middle-class audiences were reached, the founders of the New England Kitchen constructed a distinctly moralized class hierarchy. One described the poor as unreachable and "incorrigible" and blamed the problems of the public kitchens on the "extreme slowness with which the mass of people change their habits with regard to food." Educated middle-class eaters were described as willing, flexible eaters who will "taste a new dish with readiness" in contrast to stubborn and irresponsible "unhealthy others," such as the "factory worker or the average school girl who cannot be brought to try it.[28]

The failure of the New England Kitchen and simultaneous success of its World's Fair replica represented a pivot point for Progressive Era dietary reformers, who soon transitioned their focus to the more amenable, "intelligent" class. Framing their work in terms of "race betterment," Ellen Swallow

Richards, a chemist and leader of the domestic science movement, coined the term "euthenics" to describe the efforts she saw as preceding eugenics, focusing on the "better raised" rather than the "better born."[29] Throughout these efforts to secure social stability by eating right, nutritional concepts were used in the service of defining and delineating distinctions between the middle class and those beneath them in the social hierarchy.

Domestic scientists helped to bolster the unstable identity of the broad middle classes by promoting a concept of a "good diet" that was carefully calibrated to the income, occupation, experience, lifestyle, and the "habituated preference" of the eater. Dietary differentiation by income and occupation were fundamental to the principles of early nutrition that informed the work of domestic scientists. It was precisely the inappropriate expenditure on food relative to income that Atwater aimed to remedy through his research, which revealed that food value was determined by nutrients, not cost, and that the best diet was carefully matched to the income and energy demands of the eater. Building on this idea, domestic scientists provided detailed dietaries by income, suggesting for example evening meals for those subsisting on fifteen cents per person per day that included milk, homemade bread, butter, and stewed pears, while those spending one dollar per day included tomato soup, halibut, filet of beef with piquant sauce, potatoes, beets, sweetbreads, and ladyfingers.[30]

Following Atwater's principles, domestic scientists also mitigated class boundary confusion caused by the blurring of income differences between manual and nonmanual labor by arguing that "good diets" were different for people engaged in different occupations *and* that even when the energy demands of their work were similar, people of different classes required different diets. According to Richards, food for those involved in "brain work," such as professors, students, doctors, and teachers, should be "liberal, varied, well-cooked . . . well-flavored" and "delicately served." Manual workers such as housekeepers, janitors, nurses, and cooks, on the other hand, should be given hearty food in few courses and never served salads or soup. Furthermore, even when engaged in physical activity, middle-class eaters should be served different foods from manual laborers. Excessive physical activity required a lot of energy, but according to Richards, the "form in which the food is served is to be that to which the men are accustomed." Active youth needed more energy than did sedentary brain workers, she explained, but that did not mean that the Harvard boat crew should eat the same foods as youths in a logging camp or soldiers in the field. The soldier could be given bread, bacon, and beans, but the Harvard boat crew required "frills," such as ice cream and strawberry shortcake.[31] This approach to dietary differentiation clearly responded to the class concerns of the time, using the language of nutrition to naturalize socially constructed class distinctions.

While we no longer see dietary advice distinguished overtly by income, occupation, and "habituated preferences" as in the late nineteenth and early twentieth centuries, the legacy of this relationship between social class and nutrition science remains with us in the form of ongoing moralized binaries between good and bad eaters that are consistent with class hierarchies. In fact, the major dietary reform movements operating in the late twentieth century—alternative food and anti-obesity movements—have converged in the production of a binary between allegedly responsible upper-middle-class eaters (largely educated and white) and "unhealthy others." While soda drinking in reality cuts across class lines, those situated economically or culturally, or both, in today's upper-middle class generally aspire to be seen (and to see themselves) as thin, health-seeking, disciplined, informed, "slow food" eaters pursuing ethical alternatives to cheap, industrial convenience, in contrast to fast food–eating dupes of the industrial food system, who are fat, ignorant, and lacking in willpower. The dynamic between the healthy self and the "unhealthy other" as well as the propensity for upper-middle-class reformers to define good diets in ways that bolster their own identity as distinctly responsible people are clearly at play in the contemporary campaign against soda drinking and heightened by the simultaneous—but, I would argue, not unrelated—juicing craze among the self-same upper-middle class.

Looking through the lens of the past, it becomes clear that adulation of fresh-pressed juices and concerns about excessive soda drinking are clearly not just about nutrients. Beyond the distinctions between these two beverages that can be measured and quantified through the numerical language of nutrition lies a world of social and moral signification. The moral implications of public health campaigns aimed at reducing soda consumption are intrinsic, not incidental. Seeing and attending to the social role of dietary advice and dietary ideals does not deny the nutritional and physiological reality, but it does change our understanding of what exactly nutrition is and does. When we talk about green juices and sugary drinks, or any other "good" or "bad" foods or diets, we are inevitably also talking about social values and social ideals established by the economically and culturally dominant members of society. The dangers to the bad eater are not only medical but also social. Consequently, the moralized hierarchies produced through the language of good and bad food (and drinks) is inevitably tied to class.

NOTES

1. Thanks to the Style.com journalist whose questions about green juices got me thinking about the relationship between good juices and bad sodas. "How Wellness Became the New Luxury Status Symbol," 19 January 2015, https://www.yahoo.com/lifestyle/how-wellness-became-the-new-luxury-status-symbol-108665751463.html.

2. J. C. Louis and Harvey Z. Yazijian, *The Cola Wars: The Story of the Global Corporate Battle between the Coca-Cola Company and Pepsico, Inc.* (New York: Everest House, 1980).

3. Juicepress, 1 March 2016, https://juicepress.com/; Helena Bottemiller Evich, "Berkeley Breaks through on Soda Tax," *Politico*, November 2014, http://www.politico.com/story/2014/11/berkeley-breaks-through-on-soda-tax-112570.html.

4. John Coveney, *Food, Morals, and Meaning: The Pleasure and Anxiety of Eating* (New York: Routledge, 2000), 63.

5. Charles E. Rosenberg, *No Other Gods: On Science and American Social Thought* (Baltimore: John Hopkins University Press, 1961); Harmke Kamminga and Andrew Cunningham, "Introduction: The Science and Culture of Nutrition, 1840–1940," in *The Science and Culture of Nutrition, 1840–1940*, ed. Harmke Kamminga and Andrew Cunningham (Amsterdam: Rodopi, 1995), 13.

6. Coveney, *Food, Morals, and Meaning*, 53–56.

7. W. O. Atwater and Charles Woods, *The Chemical Composition of American Food Materials* (Washington, D.C.: U.S. Department of Agriculture, Office of Experiment Stations, 1899).

8. Atwater, "Pecuniary Economy of Food: The Chemistry of Foods and Nutrition V," *Century*, November 1887–April 1888, 437.

9. Atwater, "Pecuniary Economy of Food," 445.

10. Irma Jones, "Ethics of the Kitchen," *New England Kitchen* (December 1894): 109, 112.

11. Mary Hinman Abel, "The Story of the New England Kitchen: Part 2, A Study in Social Economics," in *The Rumford Kitchen Leaflets: Plain Words about Food*, ed. Ellen Richards (Boston: Home Science Publishing, 1899), 149, 137.

12. Maria Parloa, "The New England Kitchen," *Century* (December 1891): 317.

13. For further analysis, see chapter 4, "From Microscopes to 'Macroscopes,'" in my *Eating Right in America: The Cultural Politics of Food and Health* (Durham, N.C.: Duke University Press, 2013), 80–108.

14. See also *Eating Right in America*, 31–33; Julie A. Reuben, "Beyond Politics: Community Civics and the Redefinition of Citizenship in the Progressive Era," *History of Education Quarterly* 37, no. 4 (Winter 1997): 418–19.

15. Harvey Levenstein, *Paradox of Plenty: A Social History of Eating in Modern America* (New York: Oxford University Press, 1993), 57–58.

16. *Proceedings of the National Nutrition Conference for Defense* (Washington, D.C.: Federal Security Agency, Office of the Director of Defense, Health and Welfare Services, 1941); Food and Nutrition Board Committee on Diagnosis and Pathology of Nutritional

Deficiencies, *Inadequate Diets and Nutritional Deficiencies in the United States* (November 1943), 1–3.

17. Karen Anderson, *Wartime Women: Sex Roles, Family Relations, and the Status of Women during World War II* (Westport, Conn.: Greenwood, 1981); Amy Bentley, *Eating for Victory: Food Rationing and the Politics of Domesticity* (Urbana: University of Illinois Press, 1998); Lawrence R. Samuel, *Pledging Allegiance: American Identity and the Bond Drive of World War II* (Washington, D.C.: Smithsonian Institution Press, 1997); Allan M. Winkler, *Home Front U.S.A.: America during World War Two* (Arlington Heights, Ill.: Harlan Davidson, 1986).

18. "Food First," Cornell University radio brief, 21 August 1941. Marion Jordan Ulmer's 1942 pamphlet *Feeding Four on a Dollar a Day* explained that the knowledge of nutrition had become a "patriotic duty for everyone" and that "food can win or lose the war," "make or break the nation," and "build or destroy morale." The 19 June 1941 radio brief "Your Child's Eating Habits" from the New York State College of Agriculture Farm Radio at Cornell University suggested that "a poor diet may be the cause of lack of courage and willpower" and that mothers might guide their children "in the food habits that build health and character." The New York State Nutrition Committee published a pamphlet, *Vitality Vigor Vim Vitamins for Victory: Plan for Vitamins Every Day*, that describes people who ate a good diet as "more apt to be alert, energetic, optimistic, cooperative; they are more apt to have clear eyes, good eyesight at night, good muscles, good posture, good color, straight strong bones, and a sense of well-being" (Ithaca: New York State College of Home Economics), collection 23-2-749, box 19.

19. Clive McCay et al., *Eat Well to Work Well: The Lunch Box Should Carry a Hearty Meal*, War Emergency Bulletin 38 (1942); *Cornell Bulletin for Homemakers* 524 (1942).

20. Eric J. Oliver, in *Fat Politics: The Real Story behind America's Obesity Epidemic* (New York: Oxford University Press, 2006), reframes the obesity epidemic as an epidemic of *ideas*, and traces its emergence and virus-like spread to Dietz, whom he refers to as "patient zero." See Oliver, *Fat Politics*, 38–43, for a comprehensive overview of the role Dietz's slides played in the formation of what would become known as the obesity epidemic.

21. Reuters, "U.S. Male Soldiers Are Getting Fatter," *San Diego Union-Tribune*, 10 November 2001; Lisa Hoffman, "Half of U.S. Military Population Overweight, Study Says: A Few Extra Pounds Likely to Affect Combat Readiness of a Few Good Men," *Washington Times*, 1 January 2002.

22. *National Body Challenge*, Discovery Health Channel, 2004.

23. Robert H. Wiebe, *The Search for Order, 1877–1920* (New York: Hill and Wang, 1967), 13.

24. Stewart M. Blumin, *The Emergence of the Middle Class: Social Experience in the American City, 1760–1900* (Cambridge: Cambridge University Press, 1989), 290–96; Samuel P. Hays, *The Response to Industrialism 1885–1914*, 2nd ed. (Chicago: University of Chicago Press, 1995), 99.

25. Robert Crawford, "The Boundaries of the Self and the Unhealthy Other: Reflections on Health, Culture and AIDS," *Social Science Medicine* 38, no. 10 (1994).

26. Crawford, "The Boundaries of the Self and the Unhealthy Other," 1348.

27. Ellen Richards, ed., *The Rumford Kitchen Leaflets: Plain Words about Food* (Boston: Home Science Publishing, 1899).

28. Mary Hinman Abel, "Public Kitchens in Relation to Workingmen and the Average Housewife," in *The Rumford Kitchen Leaflets: Plain Words about Food*, ed. Ellen Richards (Boston: Home Science Publishing, 1899), 159.

29. Ellen Richards, *Euthenics: The Science of a Controllable Environment* (Boston: Whitcomb and Barrows, 1910), vii.

30. Ellen Richards, *The Cost of Food: A Study of Dietaries* (New York: J. Wiley and Sons, 1901), 138, 11.

31. Richards, *The Cost of Food*, 37–44.

Section III
Regulating Food

Section III examines another highly political debate, this one surrounding the role of government in the production, safety, and consumption of food and drink. Libertarians argue that the government should have no role at all in regulating the decisions we make for ourselves, and ideologically driven free-marketeers believe the government should have no role in the marketplace whatsoever. Centrists see a role for the government in creating and enforcing food safety rules, regulating markets to prevent monopolies and protect consumers from industrial waste, and guaranteeing a market for certain staple products like wheat, corn, and soybeans. Political progressives go beyond the centrist view and envision guaranteed nutrition for all, better working conditions for food producers, and fair compensation for the work that is done. But the question remains: What is the proper extent of government involvement in all of this?

This question is significant because too much regulation, or regulation in the wrong places, stifles innovation by both small and large producers, increases production costs and thus favors large companies over small ones, and creates regulatory capture by businesses and farmers (who are sometimes one and the same) who then feed from the trough created by government rules and subsidies. All of this hurts small producers and especially consumers who end up paying more money for what they buy, and who may well be getting fewer choices and inferior products compared to what a free or freer market might have produced. Yet history is littered with examples of market collapses and credit runs, monopolies and price gauging, unsafe food and polluted waterways, and unscrupulous business practices that hurt consumers and laborers alike. Indeed, history is also replete with examples of field, slaughterhouse, food assembly line, and restaurant workers who are paid a pittance for their labor, or who work in conditions that can, and sometimes literally do, kill them. It is not surprising that most nations have decided that some degree of government involvement in food production, safety, and consumption is necessary. The high stakes of this debate should be clear, and will require consideration of both the successes and failures of free markets and government intervention in the past. Luckily for us, history provides many examples.

7

Who Should Be Responsible for Food Safety?

Oysters as a Case Study

MATTHEW MORSE BOOKER

A century ago, confronted by the scale and danger of an industrializing food system, public health investigators and legislators remade responsibility for keeping food safe. When the journey from "farm to fork" was shorter, and more obvious, it made sense to place responsibility for food safety with those who made food and those who ate it. But in the industrial era, many steps exist in the journey, many hands touch the food, and a wide range of landscapes are implicated. The very geographic range and movement of food means exposure to hazards. Filthy and disease-ridden barns, slaughterhouses, and grocery shelves make food unsafe. So do polluted shellfish beds. A polluted river or harbor is not under the control of any one person or producer. Resolving that pollution is the stuff of political, economic, even cultural life.

Debates over food safety raged then and they rage now. A century ago, the debates centered around "pure" food. Epidemic illness and widespread suspicion of unscrupulous producers combined to create an age of fear. In the popular medical thinking of the late nineteenth century, purity was equivalent to safety. If a food lacked additives, if it was unfalsified, it was pure and therefore safe. Today's debate continues the concern for purity, but it largely ignores safety because safety is taken for granted. In developed economies, a battery of regulations and standards (not to mention trial lawyers) stand between consumers and their food. Food is washed, processed, irradiated, sterilized, and inspected. But it is often unrecognizable as an animal or vegetable. And the contents of food, from its genetic makeup to its added stabilizers and preservatives, create tremendous anxieties. In the popular thinking of our time, safety is assumed but purity is desired. The continuing question is responsibility. Before the twentieth century, producers and consumers were held responsible for the safety of the food they produced or consumed. But for more than one hundred years, federal and state agencies have been empowered to inspect, approve, and remove unsafe foods. These agents and their rules ensure far safer food than in the previous era. But they have not resolved our desires for purity, which has come to mean something much more elusive and harder to guarantee.

As the distance food traveled between farms and dinner tables grew longer in the nineteenth century, consumer concerns increased. People no longer knew where their food came from. Food preparation moved out of the hands of individuals and families into restaurants and bakeries. All phases of food production, processing, and consumption were reshaped by industrial processes. Organized wage labor and machines remade work and transport. As Peter Coclanis suggests in his essay in this volume, the Industrial Revolution was also an agricultural revolution, and it accompanied an urban revolution. As machines remade the countryside, displaced farmers crossed oceans to new homes. Overwhelmingly, they landed in cities. Between 1860 and 1920, the population of the United States shifted from overwhelmingly rural to majority urban. In this new, modern world, food took on new meanings. Concerns about food safety were never just about food. They reflected fear of filthy kitchens and uncertain refrigeration, anxiety about new immigrants, and fears of unsafe tenements and polluted urban environments. As Charlotte Biltekoff argues in her study of dietary reform in American history in this volume (chapter 6), "The history of dietary reform in the United States shows . . . that concerns about bad eating habits typically express broader social concerns and that advice about how to eat right conveys important social and moral ideals."

Faced with new kinds of uncertainties and anxieties around food, modern people began to turn to new ways of apportioning responsibility for food safety. In the past, safety had lain with producers or consumers, but increasingly citizens were demanding that governments monitor industrial producers to protect the mass of consumers. Early in the twentieth century, Americans looked in new ways at the age-old question "Who is responsible for the safety of food?" This chapter looks at events in a key moment, the cusp of the twentieth century, to illustrate how the answer changed in our recent past. This moment has much to tell us about the food safety panics and concerns of our own time.

The paradox of modern life is that we demand government protect us from an ever-greater range of risks decry excessive governmental control, but many of us decry excessive governmental control. Even fewer acknowledge this as a paradox. Either we don't know what we want or we aren't honest about it. This is quite visible in our relationship to food. For example, consider the oyster. The history of American oysters helps us see the emergence of new kinds of food safety concerns, environmental rather than adulteration, in the late nineteenth and early twentieth centuries. A nineteenth-century daily staple of rich and poor alike, by the 1950s oysters had declined to an occasional luxury in the American diet. While this chapter addresses a large topic, oysters will be used as a specific example to illuminate the question of responsibility for keeping food safe. The rise and fall of oysters can help us understand the role of fears about food safety.

This essay tries to find and explain the beginnings of this central tension in modern life. It begins with the panics over foodborne disease that terrified Americans in the late nineteenth century. At first Americans responded as they always had, blaming those who made or sold impure food and expecting buyers to beware. But spurred by repeated catastrophe and public outcry, federal legislators thought it politically prudent, even morally necessary, to intervene by creating new organizations to monitor and control—to *regulate*—the purity and safety of food. Over the past century, these agencies have expanded their authority, placing their stamp of approval on meat, milk, fruits, vegetables, and all forms of processed foods.

To understand the urgency of the regulators at the turn of the twentieth century, let us recall the terrible dangers they faced. Typhoid fever is long gone from the memory of most Americans, though elsewhere in the world it continues to kill roughly a million people per year, most of them children. Typhoid tortures before it kills. One of the great truths of history is that forgetting can be a blessing. Listen to a French doctor describe losing his newlywed wife: "Typhoid fever? It is the memory of atrocious suffering, of interminable nights without sleep, of the most painful fever and agitation, of nervous problems which make patients feel themselves odious to loved ones. . . . And oh! it was such a jolly dinner at which those shellfish were served . . ."[1] Typhoid's destructiveness is hard to remember in a society that suffers only a handful of cases each year. Its defeat is one of the great successes of American public health. That defeat began with the medical mystery at Wesleyan University in 1894.

FOOD FEARS IN THE AGE OF DIARRHEA

On 12 October 1894, fraternity brothers at Wesleyan University in Middletown, Connecticut, sat down to their annual pledge dinners. This was a celebratory meal on an autumn evening. We can assume the meal was a happy one, featuring toasts, laughter, and traditional foods to celebrate welcoming new members into a community. But the happiness was short-lived. A week later several students came down with a slight fever. The number of cases increased, symptoms grew more severe, and after a week it became clear that the sick were suffering from the well-known scourge, typhoid fever. By 1 November more than twenty students were ill. By the middle of November, twenty-five had sickened, thirteen very seriously, and after terrible suffering, four died. To die in the prime of life is horrible for anyone. But these were the children of the elite. They were supposed to be protected by their wealth and status. Typhoid was thought to be a disease of the crowd, of immigrants, of the urban poor. But at Wesleyan, in 1894, it killed the rich too.[2]

The Wesleyan case came in the midst of a crushing economic depression. In addition, 1894 is memorable for strikes, brutal crackdowns, and a growing fear of open class warfare. At a time when fewer than 5% of eighteen-year-olds attended university, Wesleyan's elite children were mostly insulated from the economic crisis and violence wracking the nation's cities. The illness that struck down these wealthy elites belied that sense of safety and connected the students and their families to the sufferings of other Americans in an age of disease. As with millions of others who sickened and died in the United States that year, the food the fraternity brothers ate exposed them to danger.

The fear and pain felt in Middletown in 1894 were not new in human history. Every human generation has wondered, to some degree, Is my food going to make me sick? Previous societies used laws like Hammurabi's Codes to try to impose some order on a fundamentally uncertain aspect of life. But if concern for the safety of food has been a constant, at certain times in recorded history the question of food safety has been much more urgent. In the United States, as in other parts of the industrial world, a critical moment occurred when mechanization, urbanization, and immigration coincided to create distinctly modern ways of producing and consuming food.

As health authorities, including Wesleyan biology professor H. W. Conn, began investigating the outbreak they noticed some mysterious

TABLE 3. SOME TYPHOID EPIDEMICS BLAMED ON AMERICAN OYSTERS

Year	City
1853	Charleston, S.C.
1854	New York City
1894	Middletown, Conn.
1896	Port Richmond, N.Y.
1902	Emsworth, England
1903	Orange, N.J.
1904	New York City
1905	Brooklyn, N.Y.
1906	London, England
1906	Middletown, Conn.
1907	Dorr, Mich.
1908	Frankfurt, Germany
1909	Toledo, Ohio
1909	New York City
1910	New York City
1911	New York City
1911	Goshen, N.Y.
1912	New York City
1913	New York City
1914	Naples, Italy
1917	Staten Island, N.Y.
1924	Chicago
1924	Washington, D.C.
1924	New York City

patterns. None of the infected were women. All were members of fraternities, and all had eaten pledge dinners on 12 October. At the dinners, they ate oysters, because seemingly every important nineteenth-century dinner had oysters. And the fraternity brothers ate their oysters raw, because eating raw oysters was one way wealthy young men exhibited their masculinity in 1894. Of all the possible sources of contamination, Professor Conn and his fellow investigators rapidly turned their attention to the oysters, because shellfish were implicated in numerous disease outbreaks from the middle of the nineteenth century (see table 3).

The Wesleyan typhoid outbreak was only one of many tragedies that contributed to popular demands for regulation of food at the turn of the twentieth century. Congress considered and rejected more than a hundred food bills between 1879 and 1905. Ultimately, however, Congress passed the Pure Food and Drug Act in 1906. The act, with its companion meat inspection bill passed on the same day, established the first comprehensive legislation governing food and drug safety in American history. The Pure Food and Drug Act, with updates in the 1920s, 1930s, and 1970s, remains the basis for all food and drug regulation in the United States today. The initial act placed responsibility for animal and plant inspection in the hands of the U.S. Department of Agriculture (USDA). Later modifications created and placed responsibility for approval of new foods and drugs in the Food and Drug Administration (FDA) and oversight of toxic and environmental threats to health in the Environmental Protection Agency (EPA). This triumvirate today inspects fresh produce and meat, approves new drugs for human and animal use, and sets standards for pesticides and factory emissions. Together they are responsible for assuring the safety of every aspect of food safety from meat freshness to genetically modified crops.[3]

This brief summary of the 1906 Pure Food and Drug Act and its administrative offspring is familiar to generations of lawyers, food activists, and companies seeking the blessing of federal regulators. Depending on their fate in the process, the longevity of this basic framework is either an example of its remarkable flexibility and adaptability, or a consequence of Congress's failure to update the regulatory system for newer problems. Proponents of the latter view point to a fateful Reagan administration decision not to request a new agency or regulatory agreement in the face of unprecedented challenge of regulating crops and drugs created by new techniques of genetic engineering. How should society address a new era in which the value lies not in the product but the process? How should we regulate a scramble for intellectual property that now dominates patent applications? How do we ensure the safety of materials we put into our bodies created in labs and factories as much as fields? These debates over the broad categories of biotechnology, genetically modified organisms (GMOs), synthetic biology, and personalized medicine challenge the basic division into crops and livestock (U.S. Department of Agriculture), food and medicine (Food and Drug Administration), and pesticides and environment (Environmental Protection Agency).[4]

Faced with contemporary debates over food safety and a long-standing regulatory framework, lawyers and policy makers often divorce the 1906 acts from their context. Their interest in the past is largely driven by how it restrains or enables actors in the present. The trouble with this approach is that it ignores the initial motivations for regulatory action and casts the government as unnecessarily intrusive. The past becomes a flat landscape featuring cartoon characters whose complicated reasons and contradictions disappear. That is a mistake because those contradictions are built into the laws and institutions previous generations left behind.[5] In short, those who are involved in policy making today need to appreciate the motives of past actors, lest they fail to build on what were once genuine advances.

Having said that, allow me take a small bite out of that full plate by returning to Connecticut in 1894. What can this episode and dozens like it in the late nineteenth and early twentieth centuries tell us about the era's assumptions about responsibility for food safety? What did Americans at that time see as dangerous, and what, therefore, the solution? And finally, how did those concerns and solutions shape the anxieties of our own time?

When considering the evolution of fear of food, consider how it compares to other fears. Geographer Yi Fu-Tuan divides fear into two components: alarm and anxiety.

> Alarm is triggered by an unobtrusive event in the environment, and the animal's instinctive response is to combat it or run. Anxiety on the other hand is a diffuse sense of dread and presupposes the opportunity to anticipate. It commonly occurs when an animal is in a strange and disorienting milieu, separated from the supporting objects and figures of its home ground. Anxiety is a presentiment of danger when nothing in the immediate surroundings can be pinpointed as dangerous. The need for decisive action is checked by the lack of any specific, circumventable threat.[6]

Tuan's description of anxiety aptly describes conditions created by the conjunction of urbanization, industrialization, and migration at the turn of the twentieth century. These forces created a kind of second Columbian Exchange that brought together ancient human plagues like cholera and typhoid with populations from all over the world in a new kind of living space, the industrial city. Between 1870 and 1930 the U.S. population grew by 83 million people, of whom 23 million were immigrants. During the same period, the percentage of Americans living in cities rose from 26% in 1870 to 56% in 1930. Those identified as farmers dropped from 53% in 1870 to just 22% in 1930. To put this demographic revolution in the starkest terms, during the course of a single lifetime, a nation of farmers became a nation of industrial, often immigrant urban workers (see table 4).

Such fundamental change, so quickly, had so many important consequences that a century later, historians have written hundreds of books about this period in American history. A central theme of these histories is the anxiety felt by contemporaries. This was an age of fear: fear of scarcity (timber

TABLE 4. U.S. POPULATION, PERCENT INCREASE FROM PREVIOUS DECADE, AND PERCENT FARMERS (ROUNDED)

Year	Population	Percent Increase	Percent Farmers
1870	39,818,000	27	53
1880	50,156,000	26	52
1890	62,948,000	26	42
1900	75,995,000	21	40
1910	91,972,000	21	31
1920	105,711,000	15	26
1930	122,775,000	16	22
2010	308,746,000		2

Source: U.S. Census Bureau.

famine, currency shortages); of childhood suffering (infant mortality); of world war; of natural catastrophe (dust bowls, floods, fires, the great earthquake); of confidence men and sharpers; of corporate monopoly power; of drugs and alcohol; of foreign ideas and terrorism; and of political corruption. But all these fears, the one most visceral and yet perhaps hardest to appreciate today, is disease. Indeed, the decades between 1870 and 1930 could be called the "Age of Diarrhea."

Scholars have long known that reuniting the separated families of mankind unleashed a firestorm of disease on indigenous peoples around the world. Asians, Africans, and Europeans through long contact with domesticated species and with one another had developed partial immunity and resistance to highly contagious diseases like smallpox, influenza, and measles. When those diseases traveled to the Americas with traders, soldiers, settlers, and enslaved people, they burned through native populations, leading to some of history's highest death tolls.[7] In the 1870s, as new waves of migrants from far-flung parts of the earth packed into industrializing coastal American cities, they once again created conditions for epidemic disease. In the overcrowded tenements and overwhelmed sewers of these cities, contagion flourished. Most spectacular were diseases passed from person to person, particularly through what two authors dryly call "the fecal-oral route." Cholera, typhoid fever, and other gastrointestinal diseases ravaged city dwellers, especially those with the weakest immune systems. This was a phenomenon of cities everywhere: In France, the annual death rate from typhoid alone, as late as 1930, was four per 100,000 of population; in Italy, eleven per 100,000.[8] Together with tuberculosis and influenza (the other major diseases of the crowd), diarrhea and gastrointestinal infections caused nearly half of all American deaths in 1900.[9]

The anxieties of the industrial city were made worse by the helplessness of the sufferers. Workers did what they could to improve their living and working conditions. But the risks of that society were enormous and unevenly borne. Being struck by a train or drinking bad water were simply part of life. Some risks, however, were more personally felt. Eating was particularly significant, since everyone had at least some control about what they put in their mouths. And one of the foods that many urban people ate, on a daily basis, was oysters.

By 1860, two oyster industries existed in the United States: one served the rich, and one served the poor. Elites ate high-end oysters, derived from specific locations ("Blue Points," "Rockaways") with something akin to *terroir*.[10] These were famous oysters, known for their flavor and freshness, and they demanded prices to match. These oysters were a luxury food, the filet mignon of shellfish, a must for celebrations like the one at Wesleyan in 1894.[11] But oysters also had a category akin to ground beef. "Southern" oysters, as they were called in the trade, were a hodgepodge of oysters gathered from multiple natural beds in the East, often from the Chesapeake, with significant additions from privately owned aquaculture beds. This category blended wild and farmed in ways that challenge both terms.

In fact, the line between "natural" and "cultural" has never been murkier than in the oyster industry. From at least the 1810s oyster growers regularly ferried adult oysters from Virginia and as far as Florida, stored them live on beds in Long Island Sound between New York and Connecticut, and resold them into the urban markets not only of coastal cities but anywhere water would carry them: up the Hudson River and along the Erie Canal. By the 1850s a handful of oyster growers in Connecticut and oyster dealers in New York had created an integrated production, distribution, and marketing system. Growers in Connecticut grew adult oysters to maturity, collected the annual spawn in specially selected and prepared "beds" often amended with gravel or broken shell, then sold the tiny "seed" oysters to other growers to raise to marketable size in the nutrient-rich waters off rapidly growing cities. When the trains came, oysters traveled to be eaten in Chicago, Minneapolis, and Salt Lake City, and, after 1879, to be seed oysters in new production areas in San Francisco Bay, Puget Sound, and Southern California. Eventually, at the industry's height around 1900, Long Island Sound was the nexus of a global aquaculture system that shipped millions of live shellfish by sea to European markets and that colonized new waters as far afield as Honolulu and Chile. These oysters were not just an elite luxury but a food of the urban working poor, the Big Mac of the Victorian era.[12]

All of this astonishing productivity rested on shellfish biology. Oysters filter huge quantities of water through their gills, straining out tiny particles of sediment and detritus. They digest the matter, convert it into highly nutritious, high-protein meat, and expel clear water and any grit. Because they are creatures of the tidal zone, oysters are also capable of enduring great extremes of cold and heat, wet and dry. When the tide goes out, oysters close their shells and survive on the moisture trapped within. Similarly, in near-freezing temperatures, oysters can enter a kind of suspended animation. Growers discovered that oysters can be stored for more than a week out of water with ice and simple insulation in wet straw or seaweed. These adaptations to the rich but stressful environment of the tides made oysters ideally suited to long-distance transport in the age before refrigeration. It meant that oysters could be moved by humans, not just by currents. And it meant that oysters were perfectly suited to feed on the waters of coastal cities, fertilized by huge quantities of untreated human and animal waste. Oysters ate the city, and the city ate oysters.[13]

This tight coupling of urban and food systems is surprising to us today, and while it was a brilliant adaptation to new conditions, it also carried new risks. Oysters filter everything that flows through their gills, including the living bacteria from the guts of creatures upstream. Upstream were forests, farms, dairies, but also hospitals, garbage heaps, factories. Downstream, oysters turned waste into meat, but they also completed a cycle between host and victim. Unharmed by the bacteria they harbored, oysters did not discriminate between the good filth and the nasty filth.

By the 1890s, germ theory was slowly replacing older ideas of the environmental and behavioral causes of disease. In the United States and Britain, germ theory finally explained oyster panics like that at Wesleyan. As medical historian Nancy Tomes has shown, the "rules" that came with the new germ theory were often readily accepted because they tended to parallel the rules of traditional hygiene. Germ theory both "[justified] widely accepted precautions of ventilation, disinfection, isolation of the ill, and general cleanliness" and "bestowed germicidal rationales on already trusted strategies of protection."[14] Yet some of the new sanitary science actually increased the threat of disease. Flush toilets and sewer systems removed waste from homes but also increased the disease-carrying bacteria flowing into urban waters where oysters were grown. Oysters provided some of the key evidence linking human waste to the terrible scourges of typhoid fever and cholera.

After the outbreak at Wesleyan in 1894, an examiner for the Connecticut State Board of Health carefully reconstructed the event, interviewing witnesses and cooks, tracking all the foods served, and, in a nifty bit of detective work, pinpointing the sick person whose private sewer carried bacteria into the Quinnipiac estuary, just upstream from where a New Haven oyster dealer soaked his oysters before delivery to Wesleyan's fraternities and sororities. H. W. Conn's discovery directly and incontrovertibly linked oysters to typhoid, setting off a worldwide alarm over the risk from American oysters.[15]

Traumatic episodes like the Wesleyan case were common in the late nineteenth century.[16] Food poisoning, food fraud, and dangerous foods spurred state legislatures to consider many local laws. As the oyster industry indicates, the food system was already national and even global, and thus food safety was not a matter for states alone. Yet not until 1906 did Congress give the U.S. Department of Agriculture the right to regulate food safety, set standards of purity, and, more importantly, create an enforcement mechanism located in the USDA's Bureau of Chemistry. Historian James Harvey Young painstakingly reconstructed the path to the 1906 bill. Young found that Congress had acted to protect food safety only twice before. The first U.S. food safety legislation in 1848 responded to anger at ineffective and fake drugs given to sick soldiers during the war with Mexico. That conflict was the deadliest per capita in U.S. history. For every 1,000 American soldiers who served, 110 died, overwhelmingly from yellow fever, diarrhea, and cholera—waterborne diseases of the camp. Seven times more U.S. soldiers died from disease than from combat. Following the war, public opinion blamed fraudulent medicines as aggravating factors, and Congress banned "adulterated" patent medicines, most of which had been imported from Britain.[17]

Historian James Young points to the key role of British legislation in shaping the U.S. debate. From their first law in 1860 (the Food Adulteration Act) to the Sale of Food and Drugs Act of 1875, the British Parliament had considered food and drugs as a single issue; similarly, reformers in the United States paired the two. Young argues that it was expedient for Congress to legislate against impure drugs after troops once again died in large numbers from disease during the Spanish-American War in 1898. But he and other scholars also point to persistent business concerns about British and German bans on poor-quality imported food from the United States.[18]

Food safety in the age of industrialization was a shared problem in industrial nations around the world. One of the "intellectual brokers" who carried ideas across the Atlantic was the British chemist Arthur Hill Hassall. Hassall had influenced the foundational 1860 British law that later provided a model for a series of food safety bills promoted by U.S. grocers and food manufacturers. These businessmen sought to ensure consistent quality to benefit trade more than they worried about health threats.[19] Industry groups sponsored competing legislation to punish or exclude competitors, and a handful of consumer-oriented bills also made it into Congress in the late nineteenth century. Yet almost every one of the more than one hundred food purity bills introduced into Congress between 1879 and 1906 failed. Narrow coalitions, competing interests, the extreme partisanship of the era, and especially fear of an overly powerful government led all but two of those bills to fail. The exceptions were bills to regulate glucose (corn syrup) and ersatz butter made from pig fat or plant oils, "oleomargarine." Both glucose and margarine were new products of scientific experimentation. What explains the success at regulating them was that each acted as a substitute for staple foods (cane sugar and butter), and each had powerful enemies in the producers of those staples.[20]

The 1906 act, however, was much more expansive than any previous legislation. Its purpose was to regulate dangerous industrial practices, such as those blamed for exposing oysters to typhoid, and therefore infecting oyster eaters with typhoid, as had happened at Wesleyan and places in and around New York City since 1894 (see chart of typhoid outbreaks). A consistent problem for food regulators was the power of industries to lobby for changes beneficial to their interests. One of the first to see this threat was Harvey Washington Wiley, head of the U.S. Department of Agriculture's Bureau of Chemistry. Wiley had been one of the key figures in promoting the 1906 Pure Food law, and enforcement was entrusted to his Bureau of Chemistry.[21]

In 1911, Harvey Wiley's enforcement agents targeted "floating," by which oysters were moved from saltwater growing areas to brackish areas near the mouths of creeks and rivers. Oystermen "floated" oysters just before sale so they would exchange saltwater for fresh, plumping up and also shedding some of the mud and encrusted saltwater algae. The New Haven oyster grower who sold his oysters to Wesleyan fraternities moved his oysters into the Quinnipiac River in 1894, where they were exposed to typhoid from a nearby sewer. In fall 1911 the board ruled that floated oysters were adulterated and banned storing oysters in waters of less saline content from which they are taken.[22] Over the next decade, USDA inspectors condemned shipments by 109 oyster producers or sellers, some for the presence of bacteria and others for water added to cans of processed oysters.[23] But in 1927, after industry complaints that regulations were unnecessary, the USDA considered removing its ban on floating. Now employed at *Good Housekeeping* magazine, Wiley warned of the threat to consumers. Wiley wrote, "The administration of the food law is gradually being transferred to manufacturers of food products."

And, he continued, "Here is a great industry which had been saved from practical destruction by the original ruling of the Department that no water of any kind should be added to oysters in shipment or otherwise." Now oyster producers were trying to overthrow the practices that had not only protected consumers but also saved the industry! As Wiley angrily remarked: "This is a complete surrendering to the industry of the task of making rules and regulations for conducting the industry, not in the interest of the consumer but in the interest of the producer."

Wiley's concerns reflected the changing landscape of responsibility for food safety. Producers, he contended, were willing to risk consumers' lives to make profits. The only thing standing between the companies and the public were government agencies. Guaranteeing the safety of food was a job for society's representatives at the USDA, not unwitting individuals or companies whose primary goal was profit. Wiley concluded with a prescient warning. "The Food and Drugs Act," he wrote, "was based on commercial practices which were detrimental and injurious to the consuming public. If the oyster industry is permitted to make its own regulations and its own scientific investigations there is no reason to believe that all other industries will in the near future be accorded the same privilege."[24]

Wiley won on floating oysters. The board kept its ban on storing live oysters in polluted waters. But neither consumers nor producers benefited in the long run. Faced with continuing fears of typhoid and shellfish contamination, several states closed waters to shellfish harvest, and they often chose to keep polluted waters closed rather than force polluters to clean them up. The consequences are still with us. Oyster producers lost some of their most productive urban waters, and consumers lost their appetite for oysters. Today's U.S. oyster harvest is a tiny fraction of its 1900s peak, and oysters today are a luxury food, not a working-class staple.[25]

The rise and fall of oysters captures both a shift in responsibility for food safety and an evolving understanding of "pollution." The American public of the nineteenth century demanded purity and feared adulteration. Purity was proven through chemical analysis and expressed as a percentage ("Is this flour made from 100% wheat?"). The industrial generation mistrusted producers and feared their ability to contaminate good food with false materials. Famous turn-of-the-century scandals over fake butter, meat mixed with offal, and patent drugs typify these concerns. In our own time, we too care about purity, but it is a different kind of purity. Few of us worry that our flour contains weevils or cardboard or sand, all real possibilities in 1906. We have labels on our food, and the labels inform rather than obscure what is inside the package. But for many of us, it is the naturalness and authenticity of food that are uncertain and untrustworthy.[26] The writers of the television show *Portlandia* satirized this desire with an episode on restaurant diners who demand proof of the origin of their meal. Before eating, they travel to see the farm where their chicken was raised, learning its name and quizzing the farmer about the chicken's history. This desire approaches the impossible. Americans today seem to want not only measurable purity but something like spiritual purity in their food. That kind of purity is harder to regulate, as the protracted battles over defining "organic," "local," and "natural" have shown.[27]

In Harvey Washington Wiley's day, pollution meant human and animal waste: stinking, visible, biological filth. In the age of diarrhea, waterborne diseases like typhoid and cholera carried the risk from biological pollution. The Pure Food and Drug Act of 1906 addressed the problems of the industrial city: human and animal wastes, bacterial contamination, and corrupt business practices leading to fraudulent food. Today pollution more often means toxins: invisible, cumulative dangers that are not biological in nature but chemical and environmental. This includes heavy metals, pesticides, and new cocktails of chemicals mingled in sewage treatment plants and in irrigation canals, including hormone disruptors.[28] These newer fears are products of our own time, with its own peculiar conjunction of science, technology, and food production. Our world, in the early twenty-first century, seems just as anxious and imperiled as the industrial cities of the early twentieth century. As others did a century ago, we face great change, and it is frightening. Unlike that generation, however, we inherit the legislation and institutions built to address previous food fears. Some work very well indeed. Municipal water supply, sanitary landfills, sewage treatment plants, vaccination campaigns, antibiotics, refrigeration requirements, pasteurization of milk—these multidecade investments in public health caused a fundamental shift in the age of death and cause of death of the urban population throughout the industrialized world. We live longer and healthier lives because of decisions made to address the problems of the industrial city.[29]

But some of the solutions of 1906 created unexpected problems. Regulation disproportionately benefited large companies who could afford the infrastructure to harvest, process, and refrigerate food and distribute it to the required standards. The reformers of yesteryear targeted disease-causing bacteria and biological wastes. Sterilized, plastic-wrapped, highly processed food is safe by the standards of 1906. But it can be unhealthy in ways that were not imagined. Today's Americans are threatened by chronic illnesses like diabetes, high blood pressure, and obesity, the products of too much of a good thing, "problems of plenty."[30] Our modern food system, with its vast farms and even larger corporate processors, seems hardly recognizable as "agriculture" in the sense of a century ago. This is profoundly unsettling: of all aspects of contemporary life, food seems most essentially natural and if unnatural, most potentially harmful. What to eat has become a very fraught question and has engendered its own reform movements, variously promoting vegetarianism, organic food, local food, and other alternatives to the conventional system.

The trend of people moving to the city and the factory and away from the farm that began in the nineteenth century has continued in the twenty-first. As Richard Walker notes, "Large portions of the agrarian labor process have been shifted off the farm and into the factory. The whole of the agro-production complex employs ten times the number of people as farming. While only a miniscule one-fiftieth of Americans work on farms today, over one-fifth work in the food system as a whole. . . . This is less a matter of factories *in* the fields, as Carey McWilliams called them, as of factories and fields *working together*."[31]

The food system Walker describes can seem totally removed from the world of 1894. Yet echoes of that past remain not only in the continuing relevance of agricultural production and consumer anxiety about food safety but also in the regulatory frameworks created from 1906 onward. Some of the origins of our contemporary food debates lie in the highly successful solutions of a previous time. We would do well to remember this. We would also do well to understand that the regulations of yesteryear are not adequate for the needs of today. Some laws can be discarded, and still others need to be enacted. But as always, the interests of profit, the locus of responsibility, and arguments about the proper role of government are what we are arguing about when we argue about food safety.

NOTES

1. V. M. Belin, *Coquillage set la fièvre ostréo-typhoide: Un point d'histoire contemporaine* (Paris: Presses Universitaires de France, 1934), 7, quoted in Anne Hardy, "Exorcizing Molly Malone: Typhoid and Shellfish Consumption in Urban Britain, 1860–1960," *History Workshop Journal* 55 (2003): 72–88.

2. H. W. Conn, "The Outbreak of Typhoid Fever at Wesleyan University," in Connecticut Board of Health, *Seventeenth Annual Report of the State Board of Health of the State of Connecticut, 1894* (New Haven, Conn., 1895), 243–64.

3. On the so-called coordinated framework of these three agencies, see Jennifer Kuzma, Pouya Najmaie, and Joel Larson, "Evaluating Oversight Systems for Emerging

Technologies: A Case Study of Genetically Engineered Organisms," *Journal of Law, Medicine and Ethics* 37, no. 4 (Winter 2009): 546–86.

4. Peter Barton Hutt and Richard A. Merrill, *Food and Drug Law: Cases and Materials*, 2nd ed. (New York: Foundation Press, 1991).

5. For an example of the careful, but instrumentalist legal literature, see Hutt and Merrill, *Food and Drug Law*.

6. Yi Fu-Tuan, *Landscapes of Fear* (New York: Pantheon Books, 1979), 5. See also Alison Blay-Palmer, *Food Fears: From Industrial to Sustainable Food Systems* (Hampshire, U.K.: Ashgate, 2008), 4–7.

7. Many works could be cited. As an introduction, see Alfred Crosby, *Ecological Imperialism: The Biological Expansion of Europe, 900–1900* (New York: Cambridge University Press, 1986); Charles Mann, *1491: New Revelations of the Americas before Columbus* (New York: Knopf, 2005).

8. Charles LeBaron and David Taylor, "Typhoid Fever," in *The Cambridge World History of Human Disease*, ed. Kenneth Kiple (New York: Cambridge University Press, 1993), 1071–77.

9. David S. Jones, Scott H. Podolsky, and Jeremy A. Greene, "The Burden of Disease and the Changing Task of Medicine," *New England Journal of Medicine* 366, no. 25 (21 June 2012): 2333–38.

10. We might even say *merroir*, a corruption of *terroir*, the notion that particulars of soil and climate give flavor to wine and perhaps other foods such as cheese.

11. I am currently tracking the popularity and price of these high-end oysters through one hundred years of restaurant menus at the New York Public Library's "What's on the Menu?" crowd-sourced big data project, http://menus.nypl.org/, accessed 21 March 2017.

12. Matthew Morse Booker, "Oyster Growers and Oyster Pirates in San Francisco Bay," *Pacific Historical Review* 75, no. 1 (2006): 63–88; Booker, *Down by the Bay: San Francisco's History between the Tides* (Berkeley: University of California Press, 2013); Christine Keiner, *The Oyster Question: Scientists, Watermen, and the Maryland Chesapeake Bay since 1880* (Athens: University of Georgia Press, 2010); Darin Kinsey, "'Seeding the Water as the Earth': The Epicenter and Peripheries of a Western Aquacultural Revolution," *Environmental History* 11, no. 3 (July 2006): 527–66; Jeffrey Bolster, "Opportunities in Marine Environmental History," *Environmental History* 11, no. 3 (July 2006): 567–97; Charles S. Elton, *The Ecology of Invasions by Animals and Plants* (Chicago: University of Chicago Press, 1958).

13. Booker, "Oyster Growers and Oyster Pirates," and for a visualization of the cycle from "nursery," to "feedlot" to urban market, see Gabriel Lee, Alec Norton, Andrew Robichaud, and Matthew Booker, "The Production of Space in San Francisco Bay: San Francisco Bay's Atlantic Oyster Industry, 1869–1920s," Spatial History Project, Stanford University, 15 May 2009, https://web.stanford.edu/group/spatialhistory/cgi-bin/site/viz.php?id=25&, accessed 23 November 2018.

14. Nancy Tomes, *The Gospel of Germs: Men, Women, and the Microbe in American Life* (Cambridge, Mass.: Harvard University Press, 1998), 57.

15. Hardy, "Exorcizing Molly Malone," 79–82; Conn, "Outbreak of Typhoid Fever at Wesleyan University."

16. My searches in the *New York Times*, the *Fishing Gazette*, and manuscript collections at the Whitney Museum in New Haven and Beinecke Library at Yale found oyster panics in the years described in table 3. As with all other things oyster, the first place to look is

Ernest Ingersoll, *The History and Present Condition of the Fishery Industries: The Oyster-Industry* (Washington, D.C.: Government Printing Office, 1881).

17. James Harvey Young, *Pure Food: Securing the Pure Food and Drugs Act of 1906* (Princeton, N.J.: Princeton University Press, 1989), 6–17.

18. Young, *Pure Food*, 40–52; I. D. Barkan, "Industry Invites Regulation: The Passage of the Pure Food and Drugs Act of 1906," *American Journal of Public Health* 75, no. 1 (January 1985): 18–26; Alan L. Olmstead and Paul W. Rhode, *Arresting Contagion: Science, Policy, and Conflicts over Animal Disease Control* (Cambridge, Mass.: Harvard University Press, 2015).

19. Daniel Rogers, *Atlantic Crossings: Social Politics in a Progressive Age* (Cambridge, Mass.: Harvard University Press, 2000), 4; "1875 Sale of Food and Drugs Act," http://www.legislation.gov.uk/ukpga/1875/63/enacted, accessed 1 March 2015.

20. Young, *Pure Food*, 66–94.

21. John P. Swann, "The History of the FDA," in *The Food and Drug Administration*, ed. Meredith Hickmann (New York: Nova Science, 2003), 10–11; "Harvey Washington Wiley," Chemical Heritage Foundation, last updated 10 January 2018, http://www.chemheritage.org/discover/online-resources/chemistry-in-history/themes/public-and-environmental-health/food-and-drug-safety/wiley.aspx.

22. "Oystermen Consider Needs of the Industry: They Favor Unpolluted Water and the Floating of Shellfish a Necessity," *Fishing Gazette*, 6 January 1912, 30–31.

23. Allegations ranged from "contaminated and contained enormous numbers of bacteria" (Judgment #475, 15 March 1910) to "contained added water" (Judgment #789, 1 December 1910). USDA Bureau of Chemistry, *Index of Notices of Judgment 1–10,000 under the Food and Drugs Act* (Washington, D.C.: U.S. Department of Agriculture, 1922), https://archive.org/details/usda-noticeofjudgment, accessed 20 March 2017.

24. Harvey Washington Wiley, *The History of a Crime against the Food Law* (New York: Devinne-Hallenbeck, 1929), 392–93.

25. As with seemingly everything else, New Orleans is an exception. For some of those exceptions, see Ari Kelman, *A River and Its City: The Nature of New Orleans* (Berkeley: University of California Press, 2006).

26. Kendra Smith Howard, *Pure and Modern Milk: An Environmental History since 1900* (New York: Oxford University Press, 2013).

27. Late in 2015, the Food and Drug Administration agreed to define and regulate the term "natural" on food packaging. That process is ongoing at press time.

28. Nancy Langston, *Toxic Bodies: Hormone Disruptors and the Legacy of DES* (New Haven, Conn.: Yale University Press, 2010).

29. Melosi, *Sanitary City*; Tomes, *The Gospel of Germs*; Jonathan Rees, *Refrigeration Nation: A History of Ice, Appliances, and Enterprise in America* (Baltimore: Johns Hopkins University Press, 2013); Smith-Howard, *Pure and Modern Milk*.

30. R. Douglas Hurt, *Problems of Plenty: The American Farmer in the Twentieth Century* (Chicago: Ivan R. Dee, 2002).

31. Richard Walker, *The Conquest of Bread: 150 Years of Agribusiness in California* (New York: New Press, 2004), 9.

8

U.S. Farm and Food Subsidies

A Short History of a Long Controversy

SARAH LUDINGTON

Federal support for agriculture dates back to the first president, who, himself a plantation owner, reportedly declared that he would "rather be on his farm than to be made Emperor of the world."[1] Beginning in the early days of the republic and extending through the nineteenth century, the federal government promoted agriculture by selling and giving land to settlers, promoting research and education through land-grant colleges and extension services, facilitating credit for farmers, and subsidizing water for irrigation from federal projects.[2] After its establishment in 1862, the U.S. Department of Agriculture (USDA) extended the federal support for agriculture by funding scientific research, educating farmers about new technology, and collecting and reporting agricultural statistics.[3] All that changed with the Great Depression, when, faced with a dire economic crisis, Congress authorized federal agencies to directly support farm incomes and commodity prices—and thereby to regulate the market for food. Moreover, President Franklin Roosevelt authorized the donation of federally owned surplus commodities for food assistance, linking the welfare of farmers to relief for the hungry. Roosevelt's aggressive farm and food relief policies were applauded at the time, but since then, farm subsidies and food assistance programs have been highly controversial.

The many-sided debate over federal involvement in food policy is epitomized in the reenactment of the Farm Bill, which takes place in Washington, D.C., every few years. Farmers and agribusiness (food researchers, processors, and marketers, among others) are deeply invested in maintaining the programs that channel billions of federal dollars into agriculture. Social welfare advocates are similarly invested in the Farm Bill, as it determines the funding for major food relief programs—including school lunches and food stamps. Nutritionists and public health advocates weigh in with their concerns about the availability of affordable, healthy food for the nation's poor and schoolchildren or, likewise, their concerns about the connection between inexpensive commodities like sugar and corn and health problems such as obesity and type 2 diabetes. Environmentalists want the Farm Bill to encourage conservation practices such as crop rotation, soil restoration, and organic certification. These various interest groups coalesce into a formidable bloc to support the legislation, provided that it has enough features to please each lobby. The sometimes-strange political alliances that form around the Farm Bill are an outgrowth of Depression-era legislation.

Not everyone agrees that the federal government should be spending billions of taxpayer dollars on agriculture. Free-market purists criticize the USDA for interfering in the commodities market and argue that the budget—and the commodities market—would be healthier without such spending. Others have criticized farm policies as wasteful, costing taxpayers and consumers billions and resulting in a substantial "deadweight" loss to the economy.[4] Public-choice economists cite farm lobby activity as a textbook example of rent-seeking by highly organized interest groups, and the USDA as an agency—captured by farmers and agribusiness—that seeks to expand its budget, prestige, and authority rather than contribute to the public good.[5] Small farmers and environmentalists accuse the USDA farm programs of benefiting large, commercial farms rather than small, family farms, of encouraging mono-cultivation rather than diversity, of degrading the environment and water quality, and of promoting the use of genetically modified seeds and commercial pesticides rather than more "sustainable" farming techniques.[6] The USDA has even been accused of engaging in a deliberate program of "food oppression," using its food support programs to suppress indigenous food ways and impose unhealthy eating habits on the American people in general and the poor and ethnic minorities in particular.[7] And this is just a small sample of the attacks on the USDA.

With so much criticism leveled at the USDA and its various farm and food programs, it is fair to ask what went wrong, and why. How did the New Deal legislation that Roosevelt and so many Americans considered essential to economic recovery become the focus of attack from so many different groups? And if farm policy is considered so woefully misguided, how does the Farm Bill continue to muster political support?

To understand the current debate, we need to examine the past. When we do, we see that while the major legislative changes were made in the 1930s, the seeds of current farm and food programs were sown even before the

Great Depression, in the prosperity of the 1910s and the agricultural crisis of the 1920s. The watershed Agricultural Adjustment Act of 1933 followed more than a decade of debate and experimentation with programs designed to support farm prices and incomes. In sum, these early debates shaped Depression-era agricultural policy, which in turn is the foundation of U.S. policies today.

The focus of this chapter will be on the deep history of federal farm and food programs, examining the concerns that drove the Depression-era policies and how income and price supports for farmers became closely tied to food assistance for the poor. The topic is potentially vast and extremely complicated, therefore limiting choices had to be made. I will concentrate on policies that have evolved to lie within the jurisdiction of the USDA; I will not address food safety and purity, which are largely under the supervision of the Food and Drug Administration and are addressed by Matthew Booker in chapter 7. Likewise, this chapter will not go into depth on environmental concerns, except to the extent that they specifically affected Depression-era programs.

The value of this historical look back is to understand the original motivations for policies and political alliances that have persisted, with some evolution, for the better part of a century. Without question, the original policies were intended to respond to an unprecedented economic crisis. New Deal legislation empowered the USDA to intervene aggressively in the commodities market in an effort to boost farm prices and incomes. In the process of using all means available to aid farmers, USDA acquired control over programs such as food relief and soil conservation that—while not strictly within its brief to support agriculture—were useful adjuncts to the success of its mission. The USDA put farmers first. While Congress and the USDA unapologetically attacked the agricultural crisis of the Depression in a way that favored farmers, it is overly simplistic to accuse these bodies of deliberately plotting to alter food ways, degrade national nutrition standards, or destroy family farms in the process. However, to the extent that the dire motivating circumstances of the Depression no longer exist, the uneasy alliance between farm and food policy borne of the Great Depression is arguably no longer justifiable.

The economic woes of American farmers—and the first attempts of the federal government to stabilize commodity process and farm income—began a decade before the Great Depression.[8] Prices for commodities boomed during the 1910s and then dropped precipitously after World War I ended. This collapse, known as the Agricultural Crisis of the 1920s, prompted the first attempts of the federal government to directly regulate the commodities market and influenced the way it would respond to the larger crisis of the Great Depression.

The decade before the 1920s is sometimes called the "golden age" of American agriculture. Farming was, by 1910, a largely commercial activity, meaning that the majority of American farmers were not subsistence farmers and farmers were increasingly buffeted, or buoyed, by market forces that they did not control and often did not understand.[9] In the 1910s, the market favored farmers. Farm productivity had increased, thanks to mechanized farm equipment, and demand for agricultural products was also high, thanks to an expanding domestic economy and strong international demand. From 1910 to 1914, farm prices reached their highest point since the Civil War. In addition, farm prices (the prices received by farmers) rose relative to nonfarm prices (the prices paid by farmers for nonfarm goods, such as clothing and farm machinery), causing farm incomes to rise.[10] This period of prosperity was important because it established a baseline expectation for farmers when hard times arrived; farmers sought federal assistance to regain the prosperity they enjoyed in the Golden Age, particularly regarding the ratio between farm and nonfarm prices and farm and nonfarm incomes.

Despite this rosy picture of farming, there was trouble on the horizon. Within the medical profession there was a growing interest in nutrition and the effects of malnutrition, specifically in micronutrient deficiencies that caused conditions like rickets and pellagra, and diets that caused children to be underweight, stunted in growth, susceptible to infection, or incapable of meeting the physical and mental demands of school.[11] By 1910, over four hundred public school districts had mandated annual medical examinations for schoolchildren, and by one estimate, 10% of the nation's schoolchildren showed signs of malnutrition.[12] When the draft was instituted in 1917, malnourishment came to national attention as the draft board reported that underweight (one popular measure of malnutrition) was the fifth leading cause for rejection of draftees and that 8% of rejected candidates had some kind of mental or physical defect. The Surgeon General and other public health officials publicly connected chronic childhood malnutrition with the lack of fitness for military service and deplored the nutrition-related health problems of America's children.[13] The prosperity and productivity of the nation's farmers did not translate into booming national health. Thus the stage was set.

Farmers continued to prosper during World War I. When the war started in Europe, the demand for American agricultural and manufactured products surged. By 1917, the United States was already shipping millions of tons of food to Europe to supply the allied armies, civilian populations, and the humanitarian relief efforts in Belgium and occupied France. Anticipating the need to supply its own army too, Congress passed the Lever Food Control Act of 1917, creating the Food Administration and authorizing price controls, among other tools, in an effort to stimulate production and stabilize the wartime market for commodities.[14] President Wilson appointed mining engineer Herbert Hoover, who had been coordinating the Belgian relief effort, to head the Food Administration.

The Food Administration was the first federal agency to regulate the food market, and Hoover's experience as the administrator of the Belgian Relief Corporation and the Food Administration likely had a profound effect on his later response to national hunger in the Great Depression. According to Hoover, the goals of the Food Administration were to assure an adequate food supply for the armed forces and civilians in America and Europe and to control and stabilize the distribution of food in the United States to avert problems such as panic buying, hoarding, speculation, profiteering, and food protests.[15] Hoover's regulatory strategy was twofold: to incentivize the production of key commodities such as wheat, sugar, pork, and fats, and to decrease the domestic consumption and waste of those commodities so that more could be diverted to the war effort. As befitted Hoover's personal and political philosophies, he sought to accomplish these goals with a minimum of direct federal intervention and a maximum of voluntary private action.

By statute, the Food Administration was authorized to intervene in the commodities markets, for example by setting "fair," minimum assured prices for key commodities. To stabilize prices for consumers, the Food Administration also controlled the "middlemen" involved in food processing and wholesale distribution. Hoover was especially interested in minimizing food profiteering; violent protests over the high prices of food had broken out in New York and other urban centers in the spring of 1917, showing how sudden surges in retail prices could promote civilian unrest.[16] To limit profiteering, the Food Administration established maximum profit margins for food processing and wholesale distribution, regulated the length of storage for commodities (to combat hoarding), prohibited excessive reselling, limited or banned the trading of commodities futures, and required federal permits for commodities exporters. Hoover was not authorized to set price caps on retail sales of food, and despite his efforts to secure the voluntary cooperation of retailers, consumer prices continued to rise in 1917 and 1918.[17]

Finally, given the wartime food requirements of American and its allies, and the potential for food shortages once the war was over, the Food Administration urged farmers to increase production. American farmers, pushed by Hoover and pulled by the promise of high wartime prices, complied: farmers took on debt to purchase mechanical equipment, put more acreage under cultivation, and produced record amounts of beef, pork, wheat, and corn.[18] Farmers were rewarded with the doubling of farm prices during the war, and with rising farm prices came a boom in in the value of farmland and rampant land speculation financed at high wartime interest rates. Borrowing soared: farm mortgages rose from $3.2 billion in 1910 to $10.2 billion in 1921.[19]

When the Food Administration was dissolved in 1919 after only twenty-four months in operation, it was considered to have been a successful experiment and, despite its short tenure, the legacy of the Food Administration has been significant. First and foremost, farmers learned that federal intervention in the market could be beneficial in securing a favorable price for commodities, a lesson that would soon come in handy. As the war ended and relief efforts wound down, farmers were unprepared for the rapid downward shift in demand. International demand dropped quickly as the United States discontinued its wartime credits to the former Allies and European countries revived their protectionist tariffs. Farmers were soon faced with huge surpluses, collapsing prices, and diminished income, compounded by the heavy debt burden incurred during the war and a sharp increase in railroad rates.[20] For farmers, the Great Depression had begun.

Not surprisingly, farmers turned to Congress for help, seeking relief through their three lobbying groups: the National Grange, the Farmers' Educational and Cooperative Union of America, and the American Farm Bureau Federation.[21] From 1924 to 1928, Congress considered five separate "Equality for Agriculture" bills sponsored by Senator McNary and Congressman Haugen; the first three failed in Congress, and President Coolidge vetoed the two that passed.[22] The goal of the McNary-Haugen bills was to raise and stabilize the domestic price of "basic" commodities (wheat, cotton, corn, rice, and hogs) so that farm prices would have the same purchasing power (i.e., parity) relative to nonfarm prices as they did in the Golden Age of 1910–14. To stabilize domestic prices, a government export corporation would purchase surplus commodities at the "fair" price and export them at a loss. An "equalization" fee levied on processors and distributors would cover the losses of exports dumped on the international market.[23] Critics of McNary-Haugen complained—presciently—that high prices for certain crops would stimulate greater production of those crops, that the fee levied on middlemen was an unconstitutional tax, and that other countries would raise their tariffs in response to export dumping. President Coolidge vetoed the bill in part because it privileged the farming of certain crops and would therefore promote single-crop farming.[24]

While the McNary-Haugen bills were never enacted, they show a clear connection to the basic workings of the Farm Administration and would influence the programs enacted in the New Deal. The McNary-Haugen scheme relied on a government-determined "fair" price and government purchases of grain to set a floor under commodity prices; it applied to only a few "basic" commodities, thus encouraging monoculture; and it favored farmers while visiting costs on middlemen and consumers.

By the time of the stock market crash, farmers had already endured ten difficult years. Although farm prices had slowly recovered from their 1921 low by 1929, they went into free fall in 1930 and continued to fall until 1933, as demand dried up in the domestic and international markets. The price of wheat, for example, fell from $1.00 per bushel in November 1929 to $.60 the following November and bottomed out at $.33 per bushel in November 1933.[25] Net farm revenues fell by 69% in the same time period. With banks failing, foreclosures threatened millions of farms, with almost 4% of all American farms failing in 1932.[26] In 1929, agriculture employed 25% of America's labor force[27] and also provided the country's food. It was a segment of the economy that the government could ill afford to ignore.

The Hoover administration was not indifferent to the plight of farmers. In June 1929, Hoover had fulfilled a campaign pledge to farmers by creating the Federal Farm Board. While the Farm Board was the federal government's first large-scale, peacetime intervention in the commodities market, it was not designed with economic emergencies in mind. The primary mission of the Farm Board was to strengthen agriculture by promoting the activities of agricultural co-operatives, specifically to improve their marketing organizations and thereby increase the bargaining power of farmers and boost commodity prices. The Farm Board was given a $500 million revolving fund for making loans to cooperatives and, if necessary, to form "stabilization" corporations to help manage surplus commodities. The idea was that these stabilization corporations would purchase and store commodities in times of surpluses (and low prices) and sell the stores in times of shortages, to smooth out fluctuations in prices.[28]

As fate would have it, the Farm Board was quickly overwhelmed by the economic crisis triggered by the stock market crash of 1929. In the board's first year of operation, practically every agricultural sector applied to the board for financial assistance to cope with the plummeting prices of commodities on the domestic and world markets and with a drought that threatened the livestock industry. In 1930, the Farm Board formed a Grain Stabilization Corporation to purchase millions of bushels of wheat in an effort to prop up the price. The price of wheat was stabilized, but at a low level. The board's efforts to export wheat were unsuccessful as the world market was already glutted and, after passage of the Smoot-Hawley tariff in June 1930, other countries retaliated by hiking their tariffs. By June 1931, the board was storing 257 million bushels of wheat. When it announced that it would no longer support the price of wheat, the price dropped from 52 to 36 cents per bushel in one month.[29]

Having failed in its efforts, the Farm Board ceased operations in May 1933, reporting combined losses of $300 million. Prices were lower and farmers angrier and more desperate than they had been in 1929.[30] Significantly, the board in its final report recommended that any future efforts to improve the prices of farm products should include controls on production.[31]

The Farm Board's efforts to stabilize wheat prices by storing surpluses also backfired because its massive holdings of unsalable wheat became the focal point of agitation for federal hunger relief and symbolic of the paradox of the Great Depression—of hunger amid wealth, of "breadlines knee-deep in wheat."[32] There were no federal agencies collecting statistics on hunger during the early years of the Depression, but using unemployment as a rough proxy for food insecurity, millions of Americans were suffering. By 1933, approximately 15 million people, or one-third of the workforce, were unemployed. The normal channels of relief—private charities and local government—were overwhelmed by the need and unable to supply adequate aid. With no social safety net, unemployment meant hunger for the displaced worker and his dependents.[33] Evidence exists of widespread malnutrition and deaths from starvation-related diseases.[34]

The federal government resisted providing food relief until March 1932, in part because of the widely held belief in America at the beginning of the Depression that volunteerism, charity, and neighborly concern were adequate for addressing the needs of the country's poor and unemployed. Hoover's own successes with the Committee for Belgian Relief and the Food Administration had reaffirmed his faith in voluntary, cooperative action.[35] Nevertheless, it was under Hoover that the Farm Board's surplus wheat became the impetus for the federal government's first foray into domestic food relief, despite Hoover's assertion—contrary to evidence—that "no one is going hungry."[36] By March 1931, the Farm Board was holding 150 million bushels of wheat and incurring carrying charges of about $4 million per month, and public pressure to distribute the wheat was mounting. Congress finally acted and Hoover assented; wheat distribution began in March 1932. By July 1932, the Red Cross had distributed 25 million bushels of Farm Board wheat.[37]

The presidential campaign of 1932 took place against a backdrop of increasing desperation in the agricultural sector. From 1929 to 1932, the index of farm prices had fallen 56%, whereas the index of consumer prices (the things farmers purchased for farm production and their families) had fallen only 32%.[38] Real income from farming had fallen by some 50% and farm foreclosures continued apace.[39] Farmers were angry and desperate and began to resist foreclosures by intimidating judges and auctioneers.[40] In 1932, a "Farm Holiday" movement, which was intended to be a peaceful strike by farmers, quickly devolved into forceful action. Farmers blocked highways and attacked trucks and trains trying to deliver produce; dairy farmers dumped milk onto the highways. The Farm Holiday movement dramatized the economic desperation of farmers and their growing unrest. In Congress, the president of the Farm Bureau Federation—the most conservative of the farm lobbies—warned of a pending revolution in the countryside.[41]

The revolution never materialized, but the farmers' revolt did not go unnoticed by the newly elected Roosevelt and his "brain trust." Roosevelt was convinced that boosting farm prices, which would restore the purchasing power of farmers, was essential to a general economic recovery. Thus, agricultural recovery was one of the first tasks Roosevelt tackled when he took office.[42] Roosevelt insisted that any legislation submitted to Congress must have farmers' support, so Henry Wallace, Roosevelt's secretary of agriculture, invited the leaders of the major farm organizations to Washington to help with the drafting of the new bill.[43] Farmers overwhelmingly approved the bill sent to Congress, but other interest groups who might be adversely affected by rising commodity prices—such as labor or relief workers—were not invited to consult.[44] The Farm Bill was signed into law on 12 May 1933, one day before an announced strike by the Farm Holiday movement.[45]

The Agricultural Adjustment Act of 1933 was an omnibus bill that reflected much of the experience—successful or otherwise—of the prior two decades of agricultural policy. The act had several goals: to boost farm incomes, to raise commodity prices by controlling production and removing overhanging surpluses, and to relieve farmers from crippling debt burdens. The act included multiple methods for accomplishing these ends and gave the secretary of agriculture the discretion and flexibility to use the methods he deemed most expedient.[46] The act also created a new agency within the USDA—the Agricultural Adjustment Administration (AAA)—to oversee the recovery programs.

The secretary of agriculture had two basic methods for raising farm prices. First, he could enter into marketing agreements with commodity and processing groups to control the prices paid to producers and the margins allowed by processors. This method was most commonly used with non-storable crops, such as fruit, vegetables, and milk. Alternatively, the secretary could implement a variety of supply-side controls, on the theory that market forces would cause farm prices to rise if commodity supplies shrank. In the short term, the USDA would supplement farm incomes by paying farmers to produce less. In the long term, farm incomes would improve and stabilize when prices regained their pre-Depression levels.[47] At least, that was the hope. The supply-side approach was mostly used with storable crops, such as wheat, rice, and corn (maize).

Among other powers, the act authorized the secretary to set a "fair exchange" price for commodities. This price, which later became known as the "parity" price, was defined as the price that gives a unit of a commodity the same purchasing power that it had it the "base period" of 1909–14—an idea borrowed from the McNary-Haugen bills. In other words, the act intended to return farmers to the economic status they enjoyed in the Golden Age of Agriculture, a goal that arguably created an entitlement mentality.[48]

With legislation in place, the AAA quickly set about establishing voluntary production controls for the basic crops. Using wheat as an example, farmers voted for two controls. First, farmers could contract with the AAA to receive rental payments as compensation for acreage taken out of production.[49] Second, wheat farmers voted for a "domestic allotment" plan, whereby participating farmers received per-bushel supplemental payments based on the percentage of their crop consumed domestically.[50] Participating wheat farmers agreed to cut their wheat acreage by 15% for 1934 and 10% for

1935. In return, they received adjustment payments of 30 cents per bushel on 54% of the average amount of wheat produced on the grower's farm during 1928–32.[51]

Despite the resultant crop reductions, prices for many commodities remained low because of surpluses already overhanging the market. The summer of 1933 saw continued unrest and agitation from the Farm Holiday movement. Responding to this pressure, President Roosevelt authorized the secretary to create the Commodity Credit Corporation (CCC) for the purpose of storing surplus grain and providing nonrecourse loans to farmers using the stored grain as collateral.[52] Unlike the disastrous experience of the Federal Farm Board, the CCC program was quickly judged a success. About 13% of the 1933 corn harvest was stored under CCC loans. In 1934, a drought affecting corn-growing regions caused the price of corn to rise and nearly all of the 1933 loans were repaid, with corn farmers and the CCC recording a profit.[53] The Farm Board's parting advice that successful price supports required production controls (or naturally caused scarcity) proved true.

In contrast to the success of the CCC, the AAA ran into trouble almost immediately. The 1933 farm legislation did not go into effect until after the 1933 planting and breeding season, therefore farmers faced the prospect of even greater surpluses As a result, in August 1933, the AAA announced an emergency plan to purchase and then slaughter up to 4 million piglets and pregnant sows. The campaign was a logistical nightmare and a public relations disaster. Most of the slaughtered pigs were either destroyed or converted into inedible products like fertilizer. There was an obvious and bitter irony to the destruction of perfectly good meat to raise the price of pork when millions could not afford to buy pork even at its low market price.[54]

The hog-slaughtering fiasco, and the public outrage it provoked, provided the impetus to again link agricultural surpluses with food relief. In the face of increasing public criticism, the White House in September announced a plan to distribute surplus commodities—including cured pork purchased by the AAA—as relief for the unemployed.[55] On 4 October, the Federal Surplus Relief Corporation (FSRC) was chartered for the purpose of purchasing and processing surplus agricultural commodities and distributing them for the relief of the unemployed. The FSRC firmly cemented the connection between farm support and food aid in U.S. policy.[56]

The FSRC was up and running in a month, improvising strategies to purchase, process, and distribute food. Despite distributing huge quantities of food to the unemployed, the FSRC was criticized from many sides.

Conceived in haste to resolve a public relations disaster, the FSRC would be plagued by unclear priorities and contradictory objectives.[57] To accomplish its mandate of benefiting *both* farmers and the needy, the FSRC restricted itself to distributing only commodities that were in surplus, and in amounts that were "over and above" existing relief grants. In other words, the FSRC did not want its surplus distribution to displace existing demand for food, so it avoided distributing goods when they might compete with established markets.[58]

By 1935, Roosevelt was ready to move the federal government away from providing direct relief through programs such as the FSRC. The USDA, however, still needed an outlet for its stores of surplus commodities and to avoid embarrassing displays of waste. In 1935, the USDA absorbed the FSRC, renaming it the Federal Surplus Commodities Corporation (FSCC) to better reflect its primary focus on adjusting the commodities market rather than relief. The FSCC developed even stricter distribution policies to reduce the impact of surplus distributions on existing markets. Furthermore, the FSCC left the determination of eligibility for relief entirely to the state and local relief agencies and kept no systematic information about the nature and classification of recipients. The result was wide disparities in the distribution of surplus commodities, depending on the region or state.[59]

In 1936 the FSCC also began donating surplus foods to participating school districts for free lunch programs. The USDA claimed in 1936 that it fed 350,000 children each day and sent surplus food to 60,000 schools in twenty states.[60] However, the FSCC's distribution of surplus food for school lunches was criticized for the same reasons as its distribution of food relief to unemployed. The FSCC's policies restricted the quantity and variety of commodities available to needy families and schoolchildren, causing wide fluctuations in the amounts and kinds of foods available. Commodities were purchased and donated to meet farmers' needs rather than the nutritional needs of recipients.[61]

To be sure, the USDA had supported nutrition research and policy in its Bureau of Home Economics since 1917,[62] but the FSCC, which developed its distribution policies specifically for the economic benefit of farmers, did not consider the nutritional needs of recipients in its policies. According to historian Janet Poppendieck, the housing of the FSCC wholly within the USDA was a gain for the farmer to the detriment of the hungry.[63] The institutional culture of the USDA was focused on relief of the farmer, not relief of the unemployed. Through its adjustment programs, the USDA was trying to raise farm prices by causing scarcity—thus exacerbating the very problem (the inability of needy consumers to afford food) that the surplus commodity distribution was intended to ease.[64] And without an organized lobby to look after the needs of the poor, there were no meaningful institutional checks on USDA's pro-farmer bias.

Of necessity, farm and food policy took a turn in 1936 when the Supreme Court, in *United States v. Butler*,[65] gutted the key provisions of the 1933 Agricultural Adjustment Act. Undeterred, Congress enacted new legislation to keep the USDA in the business of reducing agricultural output, this time tying acreage reduction to soil conservation. A drought lasting from 1932 until 1936 had converted a huge part of the Midwest and Plains into the notorious Dust Bowl, complete with dust storms that polluted the air in eastern cities. In 1936, Congress amended the mission of the USDA's Soil Conservation Service to include agricultural price adjustment as one of its goals. Under the soil conservation program, farmers were paid to voluntarily plant soil-enriching grasses and legumes instead of soil-depleting commercial crops.[66] Thus, the goal of increasing commodity prices became entwined with environmental conservation and food relief.

Despite the efforts of the USDA, the agricultural sector of the economy was recovering at a painfully slow pace. In 1936, farm prices and incomes had nearly recovered to their 1929 levels,[67] but it was unclear whether the efforts of the AAA were responsible for the recovery in prices or whether it was caused by years of drought and the general economic recovery.[68] When the drought ended in 1936, the USDA anticipated normal growing conditions—and huge surpluses—from the 1937 and 1938 crops. But just as the farm sector faced the prospect of farm prices dropping due to surpluses, the U.S. economy entered another recession in the autumn of 1937. Four million people lost their jobs and unemployment approached the level it had been when Roosevelt was elected.[69] Once again, local relief agencies were overwhelmed, thousands of Americans were hungry, and there were reports of people dying of malnutrition and starvation-related diseases.[70] For a second time, the country was perplexed by the problem of hunger amid plenty.

To address the new crisis, in 1938 Congress, at the urging of the farm lobby, passed a farm bill that consolidated and reauthorized existing programs and strengthened the power of the secretary to enforce supply-side controls.[71] For the first time, the secretary was authorized to set mandatory marketing quotas for corn, wheat, and rice when he determined that the crops would be in surplus. If mandatory quotas were in effect, farmers who exceeded their quotas would be fined. The quota system, working in conjunction with CCC loans, reflected the "ever-normal granary" philosophy of Secretary Wallace that agricultural policy should strive to maintain "a continuous and stable supply of agricultural commodities from domestic production adequate to meet consumer demand at prices fair to both producers and consumers."[72]

In 1939, the FSCC was invigorated by the appointment of Milo Perkins, an agricultural economist who believed that food relief was an essential tool for relieving surpluses.[73] Perkins established the first federal food stamp program in 1939. Under this program, low-income families would purchase "orange" food stamps and, for every $1 of orange stamps purchased, received 50 cents worth of "blue" stamps at no extra cost. Orange stamps could be used to purchase any food but blue stamps could only be used to purchase surplus commodities. Before it was canceled in 1943, approximately 20 million people in nearly half the counties in the United States participated in the program. However, as the food stamp program was tied to agricultural surpluses, it was terminated in 1943 because war had decimated the surpluses.[74]

Perkins was also devoted to maintaining FSCC donations to the school lunch program, which continued to receive funding for the duration of World War II even as surpluses disappeared.[75] The school lunch program was popular and had developed its own political constituency that included not only the communities served, but also the federal, state, and local officials, including nutritionists, who administered it.[76] In 1940, a school lunch lobby was formed (the Coordinating Committee on School Lunches). The continued support for the school lunch program thus depended on a coalition of social reformers, child welfare advocates, and farm and food industry lobbyists—and Perkins, who believed that the disposal of surplus commodities would be a recurring problem after the economy recovered from the war. In 1946, Congress created the National School Lunch Program, which permanently linked the "health and well-being of the Nation's children" to the growth and prosperity of the nation's farmers.[77]

By 1946, the foundations of USDA policy were firmly entrenched. Versions of the Depression-era programs—including acreage reduction programs, marketing quotas, nonrecourse loans, and government storage, purchase, and distribution of surplus commodities—persisted long past the economic crisis and can even be found in the 2014 Farm Bill. Along the way, tweaks and features were added to adjust to the complexity of the volatile world agricultural market, with a trend toward scaling back government intervention in the commodities market. In 1973, for example, Congress established relatively low "target" prices for wheat, feed grains, cotton, and rice, and made deficiency payments to eligible farmers only when the market price fell below the target price. Enacted in a time of high commodity prices and the year after the Soviet Union had purchased 440 million of bushels of American wheat, the law was designed to increase production, which would reduce the price of food for consumers and meet increasing world demand.[78] Earl Butz, the outspoken secretary of the USDA, encouraged farmers to plant "fence row to fence row" to fulfill the anticipated demand. But the notion that Butz and the 1973 legislation instituted a radical policy shift is overstated, because when farmers hit hard times, the USDA resorted to its Depression-era programs to adjust the market in favor of farmers. The boom of the 1970s went bust in the early 1980s, and in 1981 the USDA announced set-asides—production control measures that traced their lineage to the 1938 Farm Bill—for feed grains, wheat, cotton, and rice.[79]

In 1996, bowing to the demands of international trade talks, Congress did legislate a dramatic change when it finally stripped the USDA secretary of the authority to impose supply-side controls. The United States had committed itself to liberalizing agricultural trade in the Uruguay Round of GATT (General Agreement on Tariffs and Trade) negotiations, which meant the elimination of all trade barriers and programs to control commodity production. Target prices and deficiency payments were eliminated, as were mandatory set-asides. To cushion the blow of losing these supports, Congress substituted direct payments to farmers that were "decoupled" from production and scheduled to end in 2002. Known as the "Freedom to Farm Act," the law also gave farmers almost complete freedom to choose which crops to plant. Of the Depression-era programs, only nonrecourse loans and soil conservation measures remained after 1996.[80] Nevertheless, when export demand dropped and prices softened, farmers appealed to Congress, which increased and extended the direct payments.[81] American farmers either were not ready for—or unwilling to suffer—the economic buffeting of a fully unregulated market.

In fact, many farmers do not welcome the free-market approach. The agricultural market is volatile; free-market policies, and the "get big or get out" philosophy of Secretary Butz,[82] has been blamed, rightly or wrongly, for the loss of small family farms, the growth of massive corporate farms, the diversion of profits from farmers to agribusiness, the explosion in crops such as corn and soy, the rise of "factory farms" for livestock that depend on cheap feed grains, and the prevalence of fatty meats and cheap corn-based sweeteners that contribute to obesity and other health problems.[83] Despite waning government support for price supports and the idea of parity prices and incomes, many farmers—especially small ones—yearn for a return to the Golden Age of the 1910s.

While the USDA's farm policies have changed course, its food relief and conservation programs have accelerated, even though these programs respond to the concerns of constituencies not directly involved in farming. Environmental groups, for example, have increasingly inserted their agenda into the Farm Bill. While soil conservation had been part of farm policy since 1936, the environmental lobby traded its support for the 1985 Farm Bill to strengthen soil erosion programs. The Conservation Reserve Program (CRP) in the 1985 bill paid farmers to remove some 38 million acres of highly erosive farmland from agricultural production and improve the land by establishing vegetative cover.[84] The CRP has survived and even grown, despite the elimination in 1996 of other acreage reduction programs and cuts to conservation programs in 2014.

The USDA has retained jurisdiction over the two main federal food programs—school lunches and food stamps (now called SNAP, the Supplemental Nutrition Assistance Program)—and gained control of the Special Supplemental Program for Women Infants and Children (WIC) and the Temporary Emergency Food Assistance Program (TEFAP).[85] This, too, was the outgrowth of a historical process. The so-called Hunger Lobby[86] had gained traction in the 1960s, when the "discovery" of hunger in America sparked renewed interest in nutrition, the school lunch program, and food relief for the poor. Pressure from the Hunger Lobby caused Congress to revitalize the food stamp program in 1970, putting it under the management of the newly created Food and Nutrition Service (FNS) within USDA. Since 1970, the Hunger Lobby has fought, mostly effectively, to maintain food programs and protect them from budget cuts.[87]

The connection between farmer welfare and food relief remains strong. USDA food purchases, including of surplus commodities, have become an increasingly important source of supplies for food pantries and soup kitchens. In 2012, for example, over 723 million pounds of food purchased by USDA, valued at $544 million, were distributed through TEFAP.[88] However, that is only part of the story. The use of surplus commodities in school lunch programs has declined, accounting for only about 17% of the overall value of food in the lunch program by the 1990s, and surplus foods are no longer part of the food stamps program.[89] Not surprisingly, the USDA is still criticized for having a fundamental conflict of interest between its primary mission of helping the farmer and its secondary responsibility of feeding the poor.[90]

The continued housing of food and nutrition programs within the USDA reveals the persistence of the Depression-era coalition of farmers and advocates for food relief and school lunches. The USDA's food and nutrition programs are authorized in the Farm Bill and supported by vote trading between congressional representatives of the farm belt and the urban and rural poor. Depression-era farm policy established programs and created constituencies that have lasted for the better part of a century, despite the seemingly intractable conflicts of interest between the needs of farmers (for stable, relatively high food prices and a livable income) and the needs of the consuming public, especially those who depend on aid for adequate food and nutrition. Because the Farm Bill benefits both groups, the coalition of political interests has remained solid, despite concerted efforts to decouple farm supports and food programs.[91]

One example of the strength of this coalition can be seen in passage of the 2014 Farm Bill. The Tea Party branch of the Republican Party, which is ideologically opposed to both farm supports and food programs, refused even to vote on the Senate's 2012 Farm Bill. In 2013, the Senate again sent a Farm Bill to the House, proposing to cut $4 billion from the food stamp program over ten years. House Republicans proposed cuts of more than $20 billion from the food stamp program over ten years, but even this was not draconian enough for the Tea Party, which wanted $40 billion in cuts. The Tea Party was able to secure separate votes on changes to the farm subsidy and food stamp programs, but both measures passed the House without a single vote from a Democrat.

Despite its success in splitting the farm and hunger votes, the Tea Party ultimately achieved much less than it wanted. The 2014 Farm Bill that was finally enacted was a compromise measure negotiated by two Democrats and two Republicans; it cut the "farm safety net" by $8.59 billion and reduced funding for food stamps by $8 billion, rather than the $40 billion the Tea Party had wanted.[92] The bill was enacted over the objections of both liberals and conservatives to the food stamps cuts—for being too deep or not deep enough, respectively.[93] Thus, dividing farm and food interests was successful in cutting funding for both programs, but the coalition of support for the Farm Bill ultimately succeeded in dampening the effects of the Tea Party attack.

Looking back on the history of farm subsidies, perhaps the only indisputable point is that farmers have been remarkably successful politically. Despite the attacks rounded on USDA programs by economists, environmentalists, consumer advocates, small-farm advocates, and, at times, the broader public, Congress continues to spend billions of dollars on farm programs. Although the form of the subsidies flowing to farmers has changed since the Great Depression, the fundamental mission and dynamic of USDA farm policy has remained the same: the USDA uses tax dollars to support agriculture and subsidize farm incomes, for the primary benefit of farmers. Food relief, nutrition, conservation, and other ancillary programs that have been placed under the USDA umbrella are afterthoughts, included in the Farm Bill and administered by the USDA only to the extent that they can usefully support farmers.[94] Critics like Michael Pollan and Marc Bittman, who accuse the USDA of prioritizing the interests of farmers, ignore the historical point that, at least since the 1930s, helping farmers has been the USDA's mission. If the USDA is in the business of welfare, it is for the welfare of farmers, not the broader business of social welfare. This is very similar to what Peter Coclanis has to say about the history of Big Ag in chapter 2. And while one might not admire corporate farming or the USDA, one must acknowledge that both have been remarkably successful in doing what they were created to do.

The political success of farmers can be attributed to several historical accidents. The farm sector was in particular need of support when the Depression started because it had already endured a decade of low prices. Farmers were fortunate to have organized lobbies in place when the Depression hit and could advocate for advantageous policies, whereas lobbies did not exist for nutritious food assistance, environmental conservation, the preservation of small farms or food ways, or land tenancy reform. Without effective representation in Congress, these interests were either neglected by farm policy or, if they could support agricultural adjustment, were incorporated into farm policy but made subordinate to the goal of lifting commodity prices and improving farm incomes. The extent to which those goals remain subordinate is the by-product of the political process: entrenched farm and agribusiness interests, in coalition with emerging conservation and hunger lobbies (among others), have kept the Farm Bill primarily focused on serving the economic interests of farmers since 1933. There is nothing inevitable about the prioritizing of farm policy or the political coalition of farmers and food interests, but as the Tea Party's unsuccessful effort in 2013 indicated, changing those priorities and breaking that coalition will be difficult, whether the effort to break it comes from the political right or the left.

Over time, spending on the food programs administered by USDA has overtaken spending on farmers. The nutrition programs in the 2014 Farm Bill accounted for 79.1% of its cost, whereas commodity, crop insurance, and conservation programs accounted for 20%.[95] Given these funding priorities, the USDA could well be renamed the USFA—the U.S. Food Administration—but the USDA is unlikely to switch its priorities without a statutory mandate. Perhaps what is needed is a complete decoupling of farm and food policies, or a new agency devoted exclusively to the management of federal food and nutrition programs. But that change would bring its own political perils. As Charlotte Biltekoff shows in chapter 6, proper "nutrition" is an ever-moving target that follows the interests of the well-to-do. Moreover, given the realities of partisan politics, the decoupling of farm and food policies will require a massive exercise of political will—perhaps one that can only be born of a new economic crisis.

NOTES

1. Thomas Jefferson Randolph, ed., *Memoirs, Correspondence, and Private Papers of Thomas Jefferson* (London: Colburn and Bentley, 1829), 503.

2. Ronald D. Knutson, J. B. Penn, and Barry L. Flinchbaugh, *Agricultural and Food Policy*, 4th ed. (Upper Saddle River, N.J.: Prentice Hall, 1998), 245–46; Department of Agriculture Organic Act of 1862.

3. Ken A. Ingersent and A. J. Raynor, *Agricultural Policy in Western Europe and the United States* (Northampton: Edward Elgar 1999), 59; *An Act to Establish a Department of Agriculture*, U.S. Statutes at Large, 37th Congress, 2nd sess. (1862): 387–88 (available at http://memory.loc.gov/cgi-bin/ampage?collId=llsl&fileName=012/llsl012.db&recNum=418).

4. Bruce L. Gardner, *American Agriculture in the Twentieth Century, How It Flourished and What It Cost* (Cambridge, Mass.: Harvard University Press, 2002), 239.

5. Gardner, *American Agriculture in the Twentieth Century*, 245; E. C. Pasour Jr. and Randal R. Rucker, *Plowshares and Pork Barrels: The Political Economy of Agriculture* (Oakland, Calif.: Independent Institute, 2005), 53–69.

6. Gardner, *American Agriculture in the Twentieth Century*, 231, 240.

7. See, e.g., Andrea Freeman, "The Unbearable Whiteness of Milk: Food Oppression and the USDA," *University of California, Irvine, Law Review* 3 (2013): 1251.

8. Farmers are not a homogeneous group, and their interests can differ sharply based on region, the commodity they raise or grow, and their status as landowner or tenant. However, a detailed analysis of the regional and class differences among farmers is beyond

the scope of this chapter. It can safely be said, however, that the Depression-era farm subsidies did little to ease the poverty of the poorest farmers. The early policies relied on voluntary acreage reductions. Sharecroppers and tenant farmers were often displaced by those reductions or did not receive their share of the benefit payment. Thus, owners of large farms benefited more from USDA farm and food subsidies in the early twentieth century than tenant farmers. William E. Leuchtenburg, *Franklin D. Roosevelt and the New Deal: 1932–1940* (New York: HarperCollins, 1963), 137–38.

9. Ingersent and Rayner, *Agricultural Policy in Western Europe and the United States*, 56.

10. Ingersent and Rayner, *Agricultural Policy in Western Europe and the United States*, 69.

11. A. R. Ruis, "'Children with Half-Starved Bodies' and the Assessment of Malnutrition in the United States, 1890–1950," *Bulletin of the History of Medicine* 87 (2013): 381.

12. Ruis, "'Children with Half-Starved Bodies,'" 386.

13. As there was no uniform way of measuring malnourishment in the 1910s, the rates of malnutrition reported are difficult to interpret. Harvey Levenstein, *Paradox of Plenty: A Social History of Eating in Modern America* (New York: Oxford University Press, 1993), 57–58; Ruis, "'Children with Half-Starved Bodies,'" 390–91.

14. Murray R. Benedict, *Farm Policies of the United States, 1790–1950* (New York: Twentieth Century Fund, 1953), 162–64; Hugh Rockoff, *Drastic Measures: A History of Wage and Price Controls in the United States* (New York: Cambridge University Press, 1985), 50–52.

15. Herbert Hoover, introduction to *History of the United States Food Administration: 1917–19*, by William C. Mullindore (Stanford, Calif.: Stanford University Press, 1941), 1–2.

16. William Frieburger, "War, Prosperity, and Hunger: The New York Food Riots of 1917," *Labor History* 25 (1984): 217–39.

17. By the time of the Armistice, retail prices for twenty-four "principal" foods had risen about 60% from their 1913 price. Mullindore, *History of the U.S. Food Administration*, 320; Matthew W. Richardson, *The Hunger War: Food, Rations and Rationing 1914–1918* (Barnsley: Pen and Sword Books), 236; Hugh Rockoff, "US Economy in World War I," *EH.Net Encyclopedia*, ed. Robert Whaples, 10 February 2008, http://eh.net/encyclopedia/u-s-economy-in-world-war-i/. Prices rose steadily, rather than suddenly, and despite what must have been continued hardship for poor consumers, there were no more food riots after the United States entered the war.

18. "Agricultural Policy," in *The American Economy, A Historical Encyclopedia*, vol. 1, ed. Cynthia Clark Northrup (Santa Barbara, Calif.: ABC-CLIO, 2003), 328.

19. Ingersent and Rayner, *Agricultural Policy in Western Europe and the United States*, 69.

20. Gardner, *American Agriculture in the Twentieth Century*, 135; Robert S. McElvaine, *The Great Depression: America, 1929–1941* (New York: Times Books, 1993), 35–37; Janet Poppendieck, *Breadlines Knee-Deep in Wheat: Food Assistance in the Great Depression* (New Brunswick, N.J.: Rutgers University Press, 1986), 4.

21. Two of these groups had been in place since the 1890s; one, the American Farm Bureau Federation, was formed in 1910s. Benedict, *Farm Policies of the United States*, 175–77.

22. Ingersent and Rayner, *Agricultural Policy in Western Europe and the United States*, 70.

23. Ingersent and Rayner, *Agricultural Policy in Western Europe and the United States*, 70.

24. Poppendieck, *Breadlines Knee-Deep in Wheat*, 12.

25. "Table 18—Wheat, average price received by farmers, United States," U.S. Department of Agriculture, http://www.ers.usda.gov/data-products/wheat-data/wheat-data/#Historical%20Data, accessed 4 November 2016.

26. Pasour and Rucker, *Plowshares and Pork Barrels*, 88.

27. Ingersent and Rayner, *Agricultural Policy in Western Europe and the United States*, 56.

28. Benedict, *Farm Policies of the United States*, 240–41.

29. Benedict, *Farm Policies of the United States*, 262; "Table 18—Wheat, average price received by farmers, United States."

30. Benedict, *Farm Policies of the United States*, 257–63; McElvaine, *The Great Depression*, 77.

31. Ingersent and Rayner, *Agricultural Policy in Western Europe and the United States*, 71.

32. Norman Thomas, "Starve or Prosper!," *Current History*, May 1934, 135, quoted in Poppendieck, *Breadlines Knee-Deep in Wheat*, 127.

33. McElvaine, *The Great Depression*, 80; Poppendieck, *Breadlines Knee-Deep in Wheat*, 19–20.

34. Poppendieck, *Breadlines Knee-Deep in Wheat*, 32.

35. McElvaine, *The Great Depression*, 55, 58–59, 79–80.

36. Poppendieck, *Breadlines Knee-Deep in Wheat*, 41–51.

37. Poppendieck, *Breadlines Knee-Deep in Wheat*, 71–72.

38. Benedict, *Farm Policies of the United States*, 277.

39. About 45% of mortgaged farms—over a million farms—changed hands through "distress transfers" between 1930 and 1934. Benedict, *Farm Policies of the United States*, 140; Ingersent and Rayner, *Agricultural Policy in Western Europe and the United States*, 94.

40. Leuchtenburg, *Franklin D. Roosevelt and the New Deal*, 24.

41. Leuchtenburg, *Franklin D. Roosevelt and the New Deal*, 24; Poppendieck, *Breadlines Knee-Deep in Wheat*, 78–82.

42. Leuchtenburg, *Franklin D. Roosevelt and the New Deal*, 35; Poppendieck, *Breadlines Knee-Deep in Wheat*, 79–80.

43. Gardner, *American Agriculture in the Twentieth Century*, 245; Leuchtenburg, *Franklin D. Roosevelt and the New Deal*, 49; Poppendieck, *Breadlines Knee-Deep in Wheat*, 89.

44. In fact, there were no organized consumer advocacy groups to invite to the table. Benedict, *Farm Policies of the United States*, 336.

45. Leuchtenburg, *Franklin D. Roosevelt and the New Deal*, 51.

46. Ingersent and Rayner, *Agricultural Policy in Western Europe and the United States*, 95; Leuchtenburg, *Franklin D. Roosevelt and the New Deal*, 49–52.

47. McElvaine, *The Great Depression*, 148.

48. See Pasour and Rucker, *Plowshares and Pork Barrels*, 82–83, 97–100 (commenting on the inherent arbitrariness of the concept of a "fair" or "parity" price or income).

49. Base acreage was calculated by the average acreage grown in 1930–32. Ingersent and Rayner, *Agricultural Policy in Western Europe and the United States*, 96–97.

50. Under this scheme, the secretary set a national acreage allotment for crops at a level that would meet anticipated domestic and international demand at the "parity" price, and then apportioned that allotment to individual farmers based on their historical planting of the crop. Pasour and Rucker, *Plowshares and Pork Barrels*, 115.

51. Benedict, *Farm Policies of the United States*, 307; Ingersent and Rayner, *Agricultural Policy in Western Europe and the United States*, 97; U.S. Department of Agriculture, *History of Agricultural Price-Support and Adjustment Programs, 1933–84*, by Douglas Bowers, Wayne D. Rasmussen, and Gladys L. Baker, Agriculture Information Bulletin No. (AIB-485), December 1984, 6, http://www.ers.usda.gov/publications/pub-details/?pubid=41994.

52. The CCC offered "nonrecourse" loans to farmers who had signed production control contracts with the AAA. Thus, a corn farmer could store corn and get a loan from the CCC, using the corn as collateral. The loan rate per bushel of corn was the "parity" price of corn, as set by the secretary. If the market price for corn was high enough—above the parity price and CCC storage fees—the farmer could repay his loan to the CCC and sell his corn at the higher price. If the market price fell below the parity price, the farmer could repay his loan to the CCC by forfeiting his corn in CCC storage. If, as happened many times during the Depression, the parity price set by USDA was too high, the market would fail to "clear" all of the corn and the CCC would be left holding significant stores. Pasour and Rucker, *Plowshares and Pork Barrels*, 115. However, it was envisaged that these stocks could be released during times of shortage or increased demand and thus stabilize the price of the commodity over time. Ingersent and Rayner, *Agricultural Policy in Western Europe and the United States*, 98.

53. Ingersent and Rayner, *Agricultural Policy in Western Europe and the United States*, 99.

54. Leuchtenburg, *Franklin D. Roosevelt and the New Deal*, 73.

55. Poppendieck, *Breadlines Knee-Deep in Wheat*, 121.

56. Poppendieck, *Breadlines Knee-Deep in Wheat*, 131–32.

57. Poppendieck, *Breadlines Knee-Deep in Wheat*, 132.

58. Poppendieck, *Breadlines Knee-Deep in Wheat*, 136.

59. Poppendieck, *Breadlines Knee-Deep in Wheat*, 196–200, 214.

60. Susan Levine, *School Lunch Politics: The Surprising History of America's Favorite Welfare Program* (Princeton, N.J.: Princeton University Press, 2008), 44–47.

61. Levine, *School Lunch Politics*, 48–49; Poppendieck, *Breadlines Knee-Deep in Wheat*, 223.

62. Levine, *School Lunch Politics*, 34.

63. Pasour and Rucker, *Plowshares and Pork Barrels*, 205; Poppendieck, *Breadlines Knee-Deep in Wheat*, 224.

64. Poppendieck, *Breadlines Knee-Deep in Wheat*, 225.

65. *United States v. Butler*, 297 U.S. 1 (1936).

66. Zachary Cain and Stephen Lovejoy, "History and Outlook for Farm Bill Conservation Programs," *Choices* (American Agricultural Economics Association) 4 (2004): 37–38; Leuchtenburg, *Franklin D. Roosevelt and the New Deal*, 172; Soil Conservation and Domestic Allotment Act of 1936.

67. Benedict, *Farm Policies of the United States*, 365. Even before the crash, 1929 was not a particularly good year for agriculture.

68. Ingersent and Rayner, *Agricultural Policy in Western Europe and the United States*, 102; Leuchtenburg, *Franklin D. Roosevelt and the New Deal*, 77–78.

69. Poppendieck, *Breadlines Knee-Deep in Wheat*, 234.

70. Leuchtenburg, *Franklin D. Roosevelt and the New Deal*, 249; Levenstein, *Paradox of Plenty*, 54.

71. In 1937, the Supreme Court had signaled that it would adopt a more expansive view of the power of Congress to regulate production. See *NLRB v. Jones & Laughlin Steel Corp*, 301 U.S. 1, 40 (1937).

72. A surplus was declared if total supply would exceed normal supply by more than 10% (corn and rice) or 35% (wheat). Farmers who grew in excess of their quota were fined 15 cents per bushel for corn and wheat, 0.25 cent per pound of rice. The quotas became mandatory only if approved in a referendum by two-thirds of the affected farmers. Section

303, Agricultural Adjustment Act of 1938; Gardner, *American Agriculture in the Twentieth Century,* 225.

73. Poppendieck, *Breadlines Knee-Deep in Wheat,* 241.

74. USDA Food and Nutrition Service, "A Short History of SNAP," http://www.fns.usda.gov/snap/short-history-snap, accessed 4 November 2016; Levenstein, *Paradox of Plenty,* 62–63.

75. Levine, *School Lunch Politics,* 52–53.

76. Levine, *School Lunch Politics,* 50.

77. Levine, *School Lunch Politics,* 71.

78. *History of Agricultural Price-Support and Adjustment Programs,* 29–30.

79. Ingersent and Rayner, *Agricultural Policy in Western Europe and the United States,* 288.

80. Gardner, *American Agriculture in the Twentieth Century,* 216–17; Ingersent and Rayner, *Agricultural Policy in Western Europe and the United States,* 398–99; Knutson, Penn, and Flinchbaugh, *Agricultural and Food Policy,* 277–78.

81. Gardner, *American Agriculture in the Twentieth Century,* 220; David Orden and Carl Zulauf, "The Political Economy of the 2014 Farm Bill" (paper presented at the AAEA session "The 2014 Farm Bill: An Economic Post Mortem," ASSA Annual Meetings, Boston, Mass., 4 January 2015), 2–3.

82. Secretary Butz was famous for promoting large farms and robust agricultural production, but the trend toward the consolidation of farms into large—and sometimes heavily mortgaged—enterprises, the decline of small owner-operated farms, and a rise in economically unstable tenant farms, dates back to at least 1910. Benedict, *Farm Policies of the United States,* 116, 357.

83. See, e.g., "The Facts behind King Corn," National Family Farm Coalition, http://nffc.net/Learn/Fact%20Sheets/King%20Corn%20Fact%20Sheet.pdf, accessed 4 November 2016; Tom Philpott, "A Reflection on the Lasting Legacy of 1970s USDA Secretary Earl Butz," *Grist,* 8 February 2008, http://grist.org/article/the-butz-stops-here/.

84. Knutson, Penn, and Flinchbaugh, *Agricultural and Food Policy,* 273.

85. Jeffrey M. Berry, "Consumers and the Hunger Lobby," *Proceedings of the Academy of Political Science* 34 (1982): 76; Robert Gottlieb and Anupama Joshi, *Food Justice* (Cambridge, Mass.: MIT Press, 2010), 82; Levine, *School Lunch Politics,* 154, 180.

86. The major players in the Hunger Lobby are the nonprofit Community Nutrition Institute and Food Research and Action Center. Levenstein, *Paradox of Plenty,* 150.

87. Berry, "Consumers and the Hunger Lobby," 68–78; Dennis Roth, "Food Stamps: 1932–77," USDA Economic Research Service, https://pubs.nal.usda.gov/sites/pubs.nal.usda.gov/files/foodstamps.html, accessed 24 February 2019.

88. USDA, "White Paper on the Emergency Food Assistance Program (TEFAP), Final Report," August 2013, http://www.fns.usda.gov/sites/default/files/TEFAPWhitePaper.pdf, 7.

89. Levine, *School Lunch Politics,* 154, 180.

90. Nutritionist Michael Latham, quoted in Levine, *School Lunch Politics,* 111.

91. Poppendieck, *Breadlines Knee-Deep in Wheat,* 97.

92. Carl Zulauf, "2014 Farm Bill, Big Picture as Seen through Spending," *Farmdoc Daily* 4, no. 95 (22 May 2014), Department of Agricultural and Consumer Economics, University of Illinois at Urbana-Champaign, http://farmdocdaily.illinois.edu/2014/05/2014-farm-bill-the-big-picture-through-spending.html. The 2014 Farm Bill also terminated the direct payment program begun in 1996, substituting a boosted

crop insurance program for the loss of those payments, and consolidated—but did not terminate—several conservation programs.

93. Molly Ball, "How Republicans Lost the Farm," *Atlantic Monthly*, 27 January 2014, http://www.theatlantic.com/politics/archive/2014/01/how-republicans-lost-the-farm/283349; Brad Plumer, "The Farm Bill Is Up for a Final Vote Soon. Here's Why So Many People Hate It," *Washington Post*, 3 February 2014, https://www.washingtonpost.com/news/wonk/wp/2014/02/03/the-farm-bill-is-set-for-a-final-vote-soon-heres-why-so-many-people-hate-it/, accessed 4 November 2016.

94. Levine, *School Lunch Politics*, 106–10.

95. Zalouf, "Big Picture."

Section IV
Gendering Food

Section IV focuses on one of the more recent debates in the realm of food production, and that concerns the role of women, especially in their own homes. Women, as mothers, make our very first food, breast milk, but in many societies, including our own, they are often explicitly or implicitly expected to be the food producers all of their lives. Certainly many women enjoy making food for others, but not all do, and certainly many do not enjoy the expectation that cooking for others is a duty and not a choice. Like it or not, for much of human history women have been responsible for producing food for their families. But, in our own time, that expectation has begun to break down. Indeed, should we even be asking, in the twenty-first century, what women's roles in food production should be? Or, does that question only make sense if we ask the same question about men? After all, men can and do now partake in feeding infants by serving breast milk or baby formula from a bottle, and men have always been able to cook for others as well as themselves.

This debate matters because it cuts to the heart of changing gender roles and expectations. A society with fixed roles for women and men relieves itself of many difficult interpersonal and interfamily negotiations by imposing those duties without regard for people's ability, interest, or happiness. But in societies like our own where the roles of women and men are no longer frozen, how do we conduct the basic tasks of living, such as preparing food, without incessant and time-consuming negotiation? After all, someone needs to put the food on the table. No doubt the best way to move forward on this question is to reject any gender associations with cooking, but just as with the allegedly male task of being the primary breadwinner, this is easier said than done.

9

What Should Babies Eat and Whose Business Is It?

AMY BENTLEY

Here it was, the big day—my baby's first solid food. For months we had been building to this moment and I knew "what to expect," for as a first-time mother with no experience I had dutifully read numerous pregnancy and childcare manuals that provided guidance for each stage. I loved the certainty of the words on the printed page that helped to clarify great unknowns, especially during those first intense, bewildering, exhilarating days after birth. Is my baby eating enough? Is he sleeping like a "normal" newborn? A product of the late-baby-boom years, I was an infant when breast-feeding rates were at their lowest levels in U.S. history. Yet given the breast-feeding renaissance that occurred toward the end of the twentieth century, I, as a parent in the mid-1990s, embraced the practice, as did other women of the educated middle class.

The best-selling pregnancy and infant advice manual my peers and I were using at the time said to begin solid foods sometime between four and six months of age, when the baby was "ready": that is, when he or she showed interest in food, demonstrated developed swallowing reflexes, and could sit upright in the high chair. An earlier generation had begun solids much earlier, but prevailing wisdom and practice in the mid-1990s held that it was better to wait. In fact, there seemed to be an unstated assumption among women I knew that the longer one delayed introducing solids, the better. The mothers most intensely focused on infant feeding seemed to want to wait until six months rather than four. One acquaintance relayed with pride that her daughter didn't taste solid food until eight months of age.

But my baby was strong and healthy, wiry and alert, and I felt he was so active that he needed solids earlier than six months. So before my baby turned five months of age, I went to the grocery store and stared, for the first time, at the aisle of baby food products: multiple brands of infant formula in all shapes and sizes, boxes of baby cereals, and seemingly hundreds of little jars of colorful foods, most with a simple pencil sketch of a winsome baby's face on the label. I picked up and examined a jar of baby food—applesauce, as I remember. I looked at the ingredients: apples, water, ascorbic acid (for vitamin C). I looked at the price. *Hmm*, I remember thinking, *aren't those the same ingredients as in a regular jar of applesauce, except in a different size*? A look at the full-sized jars of applesauce revealed that the ingredients were the same (apples, water, ascorbic acid), but the difference in price per ounce was considerable. The baby food product cost about twice as much as the regular applesauce. I also marveled at the baby food aisle's array of feeding-related baby products: bibs, bottles, pacifiers, sippy cups, bowls, spoons, forks. Here lay an entire new world of goods, a whole niche market of products for infants.

THE BIG QUESTIONS

With the overwhelming amount of information and number of products available it's no wonder that American parents feel bewildered and anxious about infant feeding. *When do I begin feeding my baby solids*? *Should I feed my child commercial or homemade baby food*? *And which food should I feed my child first*? After the first, arguably biggest decision whether to breast- or bottle feed, these three questions become central, appearing in every mommy blog and parenting magazine and are the subject of many a Millennial dinner party. Through the long twentieth century in the United States these questions have had dramatically different answers depending on, among other things, ideas about the body and health, the current state of scientific knowledge and practice, the availability of products, and their advertising and marketing campaigns designed to create desire.

At what age do I begin to feed my child solids?

While today the generally agreed-upon age for introducing solids is between four and six months of age, the advice and practice has shifted wildly. In the space of about fifty years, from the late nineteenth to the mid-twentieth centuries, infants in the United States went from nearly exclusive consumption of breast milk in the first year, to bottle feeding of formula and beginning solids at four to six weeks postpartum—sometimes just hours after birth. This fairly dramatic change was the result of many social and economic components of

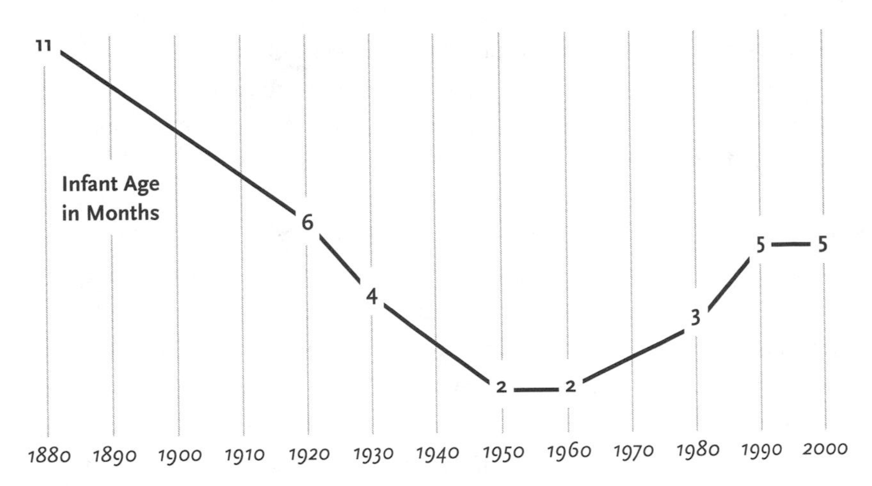

Graph 1. Commonly Accepted Age of Introducing Solids. In the space of a few decades, U.S. infant feeding norms changed dramatically—from mostly breast milk or formula substitutes through much of the first year at the turn of the century to the introduction of solids at four to six weeks by midcentury. As scientific studies revealed the negative effects of early solids, the advice and practice began to reverse with regard to age. (Graph by Ruby Gary)

the late nineteenth and early twentieth centuries: industrialization, mass production and advertising of the food supply, changing consumption patterns, the discovery and promotion of vitamins, evolving notions of the body and health, the promotion of science as the ultimate authority, and the medicalization of childbirth and infancy.[1]

In the nineteenth century, before the rise of commercial baby food, nearly all infants were breastfed exclusively for most of their first year. A pediatrician described the nineteenth-century practice as "the grandmothers' aphorism, 'only milk until the eruption of molars' (12–16 months)."[2] According to one researcher, "Milk alone was believed sufficient until the baby showed signs of failure, and often the young child's diet was confined to little more than milk until he was two years of age."[3]

Yet by the mid-twentieth century, formula feeding was the norm and solids were introduced into babies' diets at increasingly early ages, functioning not just as a supplement, but a substitute for breast milk. By the 1960s, formula was a ubiquitous presence in grocery stores and homes across the country, and the practice of raising American infants on commercially produced infant formula followed shortly by solid baby food was a natural, normalized process firmly entrenched in American culture.

Thus in just a few short decades commercial baby food emerged as the quintessential industrial product, and the age of introducing solids dropped accordingly. Although Gerber baby food got its start in the late 1920s and has since dominated the commercial baby food market, Harold Clapp of Rochester, New York, developed Clapp's Baby Foods a few years earlier. In 1921 Clapp, on the advice of doctors, developed a type of "baby soup," a combination of beef broth, vegetables, and cereal, when his wife was too ill to feed and care for their infant son. Apparently their son did so well on this mixture that Clapp teamed up with a local cannery and began producing it commercially. Gerber and other manufacturers entered the market a few years later, and thanks to efficient production and innovative marketing and advertising, Gerber quickly gained the majority of the growing market share, even during the Great Depression.[4] First producing pureed vegetables and fruits—they were termed "strained" or "sieved" at the time—producers soon added cereals and within a few years introduced chopped produce and dinner combinations for older toddlers. By the end of the 1930s it was clear that commercial baby food in the United States was here to stay.

By the post–World War II period, baby food had become a fully naturalized product thanks to widespread availability, persistent marketing campaigns, and strategic alliance with pediatricians and childcare experts. Commercially produced baby food was no longer a novelty but a necessity, even a requirement, according to conventional wisdom, for "properly" raised and nourished children, signifying the adoption of modern values of progress, efficiency, capitalism, industrialization, and reliance on scientific, expert opinion. Health professionals debated the midcentury practice of introducing solids to infants just weeks, sometimes days after birth. But for most parents in postwar America, the mainstream sentiment was not whether to use commercial baby food, but how early, which foods first, and in what quantities.

In the 1960s and early 1970s, as breast-feeding declined and formula feeding reached an all-time high, so did babies' consumption of solid food: on average seventy-two dozen jars in the first year of life. Baby food companies, who were manufacturing baby food in exactly the same way in which regular canned goods were being produced—that is, with added salt, sugar, fillers, and preservatives—came under fire as studies showed that such additives might be harmful for infants. The research, combined with a vigilant consumer movement and the public's dissatisfaction with industrialized food in general, led to the public as well as the medical establishment questioning the wisdom of early solids. Early feeding of solids became suspect and linked to future health risks, and the advice and practice shifted toward a later age of four to six months.

In the first two decades of the twenty-first century the optimal age to introduce solids was again in flux. Rules and practices that since the 1980s had seemed to be settled, straightforward, sound advice once again were up for reassessment. For example, in 2012 the American Academy of Pediatrics changed its recommendation to exclusive breast-feeding until six months (from the standard four to six months), but even well before 2012 there was a sense that the longer one waited to introduce solids, within reason, the better off babies would be. The World Health Organization had long specified that solids not be introduced until six months of age. This sense was reinforced in the 1990s and 2000s, as infant allergies were becoming a greater problem.

Health professionals, including the American Academy of Allergy, Asthma, and Immunology, began to recommend that solid foods be introduced one at a time and tested over a period of days. If a particular food caused an allergic reaction, the thinking went, it would be much easier to detect and subsequently remove from an infant's diet if parents were only adding one new food at a time. Childcare manuals of the period echoed this advice: Go slow with the introduction of food, they warned. Don't feel compelled to rush into it. Take your time with each food, and once you are completely sure that the applesauce, or cereal, or chicken you are feeding to your infant does not cause an allergy, then you can proceed to the next food. If parents adhered strictly to this advice, however, the introduction of even a basic array of fruits and vegetables might take months.

Through the 1990s and 2000s, as food allergies became more and more prevalent among children, pediatricians, and allergists began to change their perspective and advice. Introducing solids *too late*, many argued, instead of preventing allergies as health professionals had believed, may in fact do the opposite and be a factor in creating food allergies. Thus, by 2010 some pediatricians and registered dietitians began to suggest that parents *not* delay introducing solids, even commonly allergenic ones, beyond four to six months.[5] In short, researchers remained in the dark about what causes food allergies, a fact they openly admitted. A panel of experts in 2008 declared that there was insufficient evidence to support any dietary intervention to prevent food allergy beyond four to six months old. "Late" (six months or older) introduction of solids might in fact, they warned, be related to higher risk of allergies to all food.[6]

Scientific studies seemed to support the new approach. There was evidence that earlier exposure to egg, for example, can protect against an egg allergy.[7] One study found no clear evidence that maternal exposure, or early or delayed introduction of peanuts, had any impact on allergy, except for those infants deemed to have a "high risk" of allergies.[8] Another indicated that high peanut consumption early in life might protect *against* (instead of be a catalyst *for*) peanut allergies.[9] Further, animal studies suggested that a low dose of peanuts seemed to create a sensitization, whereas high doses seemed to prevent allergies. "Maybe we should go back to . . . actually introducing these foods earlier when the body has more tendency to be tolerant," remarked professor of pediatrics Dr. Amal H. Assa'ad.[10]

Not only does delaying solids not prevent allergies, professionals began to assert, but doing so affects palate development, inhibiting adventurousness in the taste, flavor, and variety of an infant's food choice and preference. "The research is really good that if you give kids one flavor for a week they are much less adventurous eaters after that," argued pediatrician Alan Greene. "Give them three flavors, and they are much more adventurous. . . . If they did have an allergy, yes, it's easier to figure out what it was. But that's a huge price to pay for a little benefit that is not going to impact most families."[11]

Further, it turns out that a baby does not arrive on the scene as a tabula rasa, as it were, with regard to culinary flavor profiles, as it is exposed to different tastes in utero. Researchers in the 1990s and 2000s found that amniotic fluid acquires flavors from the food a mother consumes, acclimating infants in utero to a variety of tastes. Because babies learn the culinary profiles of their family's food habits before birth and also through breast milk, which is similarly flavored by a mother's diet, once they begin solid food they respond favorably to the familiar flavor combinations they experienced in the womb and as a newborn.[12]

Is commercial baby food safe and healthy,
or is homemade the better alternative?

This question, which women asked government officials and manufacturers in the 1920s, when commercial baby food first became available, remains a central question for many American parents today. The short answer is that it depends on the historical and cultural context, on the range and quality of products available, on a caretaker's desire and ability to create baby food, and on the trade-off between convenience and control over ingredients and flavor profiles. While studies show that commercial baby food, much improved compared to products in the mid-twentieth century, has a protective effect for many American babies, there are advantages to homemade baby food for those with the time and inclination to prepare it.

The idea of "baby food" and "commercial baby food" arose almost simultaneously. Prior to the early twentieth century there was not an actual category of "baby food," since infants under a year generally were not fed solids. Instead, household manuals of the nineteenth century provided recipes for a general category known as "soft foods," which were deemed suitable for infants, the sick, or the elderly who needed a soft diet. The early twentieth-century perfect storm of events discussed above, including the discovery of vitamins and mass production of canned goods, created a context in which the notion of "baby food" eventually became a category of foodstuffs for most Americans.

Commercial baby food and food processing in general reached a "golden age" in the twentieth century postwar period, as new materials and methods were developed, products proliferated, and a larger percentage of household budgets was devoted to processed foods. Ever-increasing numbers of Americans became more mobile and moved to urban and suburban areas remote from food sources, and many sought more leisure time in part through labor-saving appliances in the kitchen. Heat-and-serve processed foods and canned, frozen, or otherwise packaged products helped provide Americans—women especially—the mobility and flexibility many desired. Processors included more and more additives in the products, not only to increase shelf life and prevent spoilage but also to restore vitamins and other nutrients diminished by modern processing techniques. Processing technology so pervasively permeated the American food supply that eventually "high-quality food" became synonymous with food capable of a long shelf life and low spoilage. Neither producers nor consumers considered chemical additives, either the emulsifiers and stabilizers or the seasonings, as a liability, but rather as a benign necessity if not a modern scientific asset. Processors matter-of-factly acknowledged the presence of seasonings, as exemplified by the ingredients listed on the label of the 1960s Beech-Nut baby food product, Carrots Cooked in Butter Sauce: "Ingredients: Carrots, Light Brown Sugar, Sugar, Butter, Salt, and Sufficient Water for Preparation."[13]

The result was a "canned food taste" that became familiar and palatable to most Americans over time and with increased consumption. Eventually the canned food taste, if not regarded as optimal, at least was not a strong deterrent to purchasing and consuming large quantities of industrially processed canned and frozen foods, and eventually became a preferred taste for significant portion of the population. An accounting executive in the

1960s, for example, noted that his children refused to drink fresh orange juice because it tasted "peculiar" to them. "They have grown up on frozen orange juice," he explained, "and they are convinced the frozen product is how orange juice should taste."[14] Manufacturing also transformed a food's texture and visual appearance, resulting in a soft, even mushy consistency, and often a paler color than its fresh counterpart.

Given that commercial baby food was prepared identically to other kinds of industrially prepared foods, that is, with salt, sugar, and other additives, the tastes and textures that babies experienced were similar (though in puree form) to those that adults consumed. There was no notion at the time that food for infants should be processed any differently. By the 1960s, 90% of American babies were fed highly processed cereals and prepared strained food, products that contained significant amounts of salt, sugar, and modified starches, before the end of the first month, an acclimation to the canned food taste and other industrial tastes began almost from birth.

By the mid-1970s, however, the landscape of infant food and feeding in the United States was in flux, influenced by a swirling, shifting societal backdrop, including the rise of the women's movement, environmental and consumer movements, and other movements for social justice. The Vietnam War and the Watergate scandal, meanwhile, provoked a distrust of science, government and corporate capitalism, especially on the part of the baby boomers coming into adulthood and wielding their social, economic, and political power. In this period a major shift in mindset had occurred with regard to highly processed food in general and commercial baby food in particular. Instead of newspaper articles touting the miracle of processed foods—its shelf stability, modern, scientific antiseptic production, and nutrition—as they had a generation earlier, there existed a commonly held skeptical, even critical public approach to the industry, an attitude more focused on its dangers than its positives.

In response to worries about the deficiencies of commercial baby food, homemade baby food experienced newfound popularity. Women embraced the rise of homemade baby food as a way to circumvent the status quo, opt out of commercial capitalism, or provide better health and nutrition for one's infant, often layered with a nascent mainstream feminist patina. Industry struggled to counter the bad publicity, and after several missteps, gradually began to respond to consumer demand by altering its products: discontinuing desserts and removing sugar, salt, and fillers. As ever-increasing numbers of women entered and remained in the workforce, commercial baby food remained popular. In the late twentieth century, baby food companies also began producing organic products, and new boutique baby food companies offered innovative new ingredients and flavors catering to contemporary tastes. The result is that commercial baby food is a better product today, thanks to a shift in the societal zeitgeist, vocal consumer demands, and industry responses.

A further interesting aspect of the commercial-versus-homemade debate is recent nutrition research that notes how for many American families, commercial baby food increases the variety and amount of fruits and vegetables an infant consumes, helping to delay the shift to less healthy alternatives such as French fries and sweets. A 2002 nutrition study showed that during the critical transition from baby foods to table foods, children begin to eat fewer varieties of health-protecting green, yellow, or red vegetables, and more ubiquitous, frequently less nutritious vegetables such as fried potatoes. Worse, during this transition infants and toddlers start to consume more energy-dense, nutrient-poor foods such as candy and carbonated sodas. Consumption was highest for those who transitioned from baby food to table food the earliest.[15]

Thus commercial baby food provided for these infants and toddlers a positive, protective factor, which disappeared as they transitioned to table food. Jarred baby foods, with their see-through packaging and attractive labels, feature a wide variety of fruits and vegetables, including many, such as squash, sweet potatoes, and mangoes, that most Americans do not prepare or consume regularly. The small portions allowing easy access to a broad variety of fruits and vegetables made it simple and convenient for parents to feed an array of produce to their children. A study of women receiving supplements from WIC (the federal Special Supplemental Program for Women, Infants and Children) confirms this pattern. Low-income mothers using commercial baby food gave a greater variety of fruits and vegetables to their babies than those who made their own baby food.[16] Researchers, who were not funded by any baby food manufacturer, found that WIC "infants aged 6–12 months who received commercial baby foods consumed a greater variety of fruits and vegetables, than infants who did not, characterized by a diet that was lower in white potatoes and higher in dark green and deep-yellow vegetables."[17]

While the studies provide compelling reasons for parents to use commercial baby food, many feel homemade is the better alternative, especially in regard to providing infants the freshest, least-processed tastes and textures of food, a type of palate development that some believe can lead to healthier eating later on. A 2012 survey noted that while a majority of Americans purchased baby food for their infants, 13% of U.S. families made their own baby food, and the numbers have been rising dramatically.[18] Beech-Nut researchers in early 2014 estimated that around a third of all baby food consumed is homemade.[19] This heightened interest in making one's own baby food was inspired in part by belief that homemade baby food was the purest, safest, most flavorful type available. These notions of safety and health intersected with the new "locavore" inclination to consume fresh, seasonal, locally grown food.[20]

Baby food DIY-ers come from all walks of life, including higher-income well-educated families, counterculture types, women concerned about product safety, health, and taste, and also ethnic minorities, whose family and cultural traditions place a premium on fresh, homemade fruits and vegetables.[21] For many, homemade baby food seems to embody a multitude of overlapping impulses. It assuages the anxiety brought on by modern culture and the corresponding need to minimize risk; it offers the ultimate guarantee of taste, freshness, and health; and it represents for some the ultimate expression of motherly love aside from breast-feeding.

It is not a given that homemade baby food is more nutritious than commercial, as much depends on how each is prepared, the freshness of the produce, and how long and under what conditions it has been stored. There is no question, however, that those who make their own baby food feel in control of the production process. They can feel assured of the freshness of the product and of its cost-effectiveness.[22] Taste is primary. One parent explained, "I wanted my kids to develop a taste for fresh, not jarred or canned. Using sweet potato as an example, there is a huge difference between what comes in a jar and what comes out of my oven. And I wanted my kids to get used to food in its more natural state. . . . Why shouldn't peaches taste ripe, and the sweet potatoes taste not bitter? I'll stick to making my own baby food."[23]

Further, for many, the homemade effort signifies love, specifically motherly love, given the long history of infant feeding and maternal bonding and contact. The practice of making homemade baby food, which in the mid-twentieth century was deemed old-fashioned, unscientific, and potentially less safe, in the twenty-first century now embodied the values of purity, wholesomeness, deliciousness, sacrifice, and maternal devotion. The mother-consumer who makes her own baby food expends much extra effort over that required to purchase baby food products, effort that imbues her baby food with added psychological and cultural value. It also suggests an artisanal craft ethos that can demonstrate a number of sentiments, including an idealized nostalgia for the values and lifestyle of preindustrial society. Further, homemade may also signify a type of conspicuous consumption, an elite economic and cultural status conferred by virtue of having the option of extra time and energy to make baby food from scratch.[24]

Which food first?

All societies begin feeding newborns with breast milk or a liquid equivalent, and all societies at some point move on to feeding their infants solid food. For a time there is a transition period, in the West known as weaning, as the mother gradually reduces the frequency of breast-feeding and as the ratio of liquid to solid food shifts. Cultures differ in belief about the appropriate age of introducing solids as well as also differ in the choice of first foods a child receives: it could be rice cereal, soup, congee, minced beef, butterfat, refried beans, an avocado, food first masticated by the mother, or a mixture containing the culture's signature flavor combination.[25]

In the nineteenth-century United States, popular advice manuals recommended that cereals or meats, not necessarily in that order, be introduced when teeth began to appear, between nine and twelve months of age, but only as thin gruel mixtures, broths, or juices. Although mothers fed infants "strength-producing" meats and cereals, advice manuals recommended that children not be given fruits and vegetables until two or three years of age. This was in part the result of Americans' wary attitude in general toward fruits and vegetables. Both medical opinion and folk practice in the United States were still influenced by the centuries-old Galenic theories of health and disease, which dictated that eating fruit made people, especially children, susceptible to fevers. Properties inherent in fruits and vegetables were thought to cause severe diarrhea and dysentery, especially in the summer. Given the laxative effect of fruits and vegetables if consumed in excess, it is understandable that people assumed fresh produce might contribute to diseases with symptoms that included diarrhea. Moreover, in this era before the discovery of vitamins, most people felt that fruits and vegetables provided excessive bulk and roughage and contributed little in the way of nourishment helpful to infants.[26]

As commercial baby food became entrenched in American food consumption, for most of the twentieth century and into the next, Americans collectively agreed on white rice cereal as the best first food for babies. White rice cereal was a continuation of the preindustrial practice of feeding thin cereal gruels to invalids, the elderly, or the very young. Visually and in terms of texture, it is easy to see how thinned white rice cereal looked and felt similar to the liquid breast milk or similar substitute. Bland was good, as it was thought that babies' digestive systems, as well as those of most adults, could not withstand strong spices and seasonings, which should therefore be avoided until children were much older.

White rice cereal became popular not for what it contained, but for what it did not; that is, it was regarded as the most benign of the grain cereals. In the 1930s and 1940s several manufacturers began to produce packaged baby food cereals, including corn, rice, wheat, barley, and oatmeal. By the 1960s refined white rice cereal won out. With its smooth, bland texture and easy digestibility, it became the default first food. Further, once it was fortified, rice cereal was regarded as a good source of needed iron. Finally, in the 1980s and 1990s, as concerns about infant allergies began to grow, rice cereal was considered safe with regard to allergies.

Early twenty-first-century thinking and practice, however, began to question the tradition. Is there any evidence, some asked, that infants' systems can't process spices and stronger flavors? How did doctors and child-care experts decide that infants couldn't tolerate distinct flavors? Was the belief a holdover from an earlier era, with its understanding of the digestive process based on centuries-old notions of humoral theory and the body? In the nineteenth-century United States, for example, women's digestive systems were believed to be so delicate that they were less able to digest strong flavors and also meat.[27] Bland, textureless food, the twenty-first-century revisionists argued, functions to dull the palate and help create a taste template that prefers such mild, plain food.

Research studies in the early twenty-first century revealed that a bland-tasting, textureless diet not only causes young children to develop an acclimation to and preference for such foods, but also contributes to their visual preference for "beige foods," as one researcher described them. A study in 2005 showed that babies fed plain diets featuring foods such as rusks, processed cereals, and milk were more likely to grow up to prefer similar-looking foods, such as white bread, plain pasta, potato chips, and milk. The early foods created a visual prototype of favored foods. Babies learned to prefer these foods not just because of taste but because these were what registered as "food" to them; that is, these foods reflected the shapes, colors, textures, and tastes of the food they were served, ate, and enjoyed when very young. Other, more colorful foods seemed foreign, and even "non–food like." In contrast, infants who were given a wide range of more colorful and textured fruits and vegetables grew up to show greater preference for and interest in such foods.[28]

These revisionist theories and practices are encapsulated in the "White Out movement," the twenty-first-century campaign to eliminate highly processed white rice cereal as the common first solid food for infants. Calling white rice cereal "junk food," the campaign's chief spokesperson, Dr. Alan Greene, drew on research on the physiology of taste and nutritional science to support his conclusions.[29] White rice cereal, Greene contended, has been the first solid food of choice without any kind of scientific evidence, rising to the top primarily through marketing by the baby food corporations. Such bland, highly processed foods, he argued, set the stage for acclimation to highly processed, nutritiously empty foods that are often high in fat, sugar, and salt. Critics of white rice cereal argue that there are other foods that contain more or better sources of iron, including pureed meat, as well as whole-grain cereal options.[30]

Further, Greene and others believe that early feeding of iron-fortified rice cereal was justified in part by the practice of early clamping of the umbilical cord, a procedure that doctors popularized in the early twentieth century. After birth an infant's umbilical cord pulsates for a minute or so, sending blood from the placenta into the infant. Clamping the umbilical cord immediately after birth was thought to prevent maternal hemorrhaging in addition to infant jaundice, yet early cord clamping prevents a great deal of blood from moving into the baby's body. Studies in the twenty-first century found that this extra blood expands an infant's blood volume by a third, boosting iron reserves and thus significantly decreasing the likelihood of anemia in the first six months.[31] If infants are allowed a full reserve of iron by delaying cord clamping for one to two minutes, advocates argued, healthy-term infants have no need for iron-fortified products, including white rice cereal, in the first six months.

Yet another critique of white rice cereal and other highly processed grains and sugars is that early consumption may affect infants' health not just in the short term, but throughout their lives. Studies on animals have raised concerns that infants fed these types of diets can experience a greater number of severe health problems later on, even if they abandon these diets early. The research, based on experiments with rats, concludes that diets high in fat and sugar lead to changes in the fetal brain's reward pathways, altering food preferences. Even more ominous is the idea that eating an unhealthy diet at an early age can compromise the health of one's own offspring, the next generation.[32] Thus, feeding infants food such as bland, highly processed rice cereal, its critics argue, puts them on the pathway to a lifetime of consumption of highly processed unhealthy calories, which is the cause of the rising rates of overweight, obesity, and related diseases, such as type 2 diabetes.[33]

Given the potential negatives of highly refined rice cereal, along with the current spate of creative approaches to food in general and infant food and feeding in particular, the final frontier of infant feeding in the twenty-first century may be to bypass the conventional pureed baby food stage all together. Instead, when infants are ready to move to solids—the key word being "ready"—they transition straight to the table food consumed by the rest of the family. Similar to the preindustrial practice of transitioning to solids, this approach, called "baby-led weaning," essentially advocates eliminating the "baby food" stage some argue was artificially constructed through the rise of commercial infant feeding products. Advocates of this practice describe their goal as "encouraging infant self-feeding to help a baby develop coordination, independence, chewing, and introduction to solid food."[34]

BABY FOOD IN HISTORICAL AND SOCIOCULTURAL CONTEXT

This chronicle of infant food and feeding practices in the long twentieth century describes sharp changes in advice and practice over just a few years. It is possible to make a general claim that baby food in the early twenty-first century is "better" than that in previous decades—the quality is improved, the age of introducing solids seems to have stabilized, and taste is more important—though there are still products of questionable value, and advertising making dubious health claims. Consumers through the century have been central to this improvement. The watchdog groups, consumer activists, and mothers who simply stopped buying substandard products and went out and created baby food companies, or made their own at home, were the catalysts for better commercial baby food. Perhaps, too, we have finally caught up with the dramatic changes wrought by industrial processing, are more in control of the technology, and have an enhanced scientific understanding of how to cater to infants' needs for optimal nutrition and palate development so that they can grow up to be healthy and adventurous eaters.

Yet the story of baby food should also be read as a cautionary tale, for the trajectory also reminds us of changes over time in scientific understanding, cultural imperatives and values, and feeding practices and food habits, and it is naive to assume that we have arrived at a place of finality. What we perceive today as the best practices in infant food and feeding will undoubtedly change in the future. As Charlotte Biltekoff points out in chapter 6, the "best practices"—whether best first foods, or making one's own baby food—are not merely a biomedical matter but also a reflection of social, often class-bound values. Without denying "nutritional and physiological reality," Biltekoff observes that both the medical expert's advice as well as the countercultural practice, for example, contain a "rule-bound empiricism."

Conversely, some elements of the baby food debate will remain the same. Tracey Deutsch reminds us in chapter 10 that domestic labor, particularly cooking, has always been bound up with ideas about women's roles as caregiver and family nurturer. While the landscape of food procurement has changed somewhat as more women participate in the paid workforce, the majority of the work, as well as the psychological expectations of "home making," still fall on women. The "long tradition of encouraging women to raise healthier, responsible, cohesive families and citizens by cooking," still largely remains women's responsibility. While such a charge has its rewards, to be sure, it also bears a burden, and mothers or other primary caregivers will likely continue to feel anxious about feeding their babies, including feeling pressured to demonstrate their love with the purchase of material goods. Parents, and mothers in particular, will also continue to experience tensions as they navigate the shoals of the experts' advice, the advice of friends and family, and their own instincts and preferences.

We are in the midst of a transition with regard to infant food and feeding practices, just as we are in a transition regarding the food system as a whole, or at least an awareness of the problems and a developing collective will to do something about them, as Ken Albala's and Rachel Laudan's chapters both demonstrate. Americans are expressing more interest in fruits and vegetables, and are making attempts at healthier eating. Younger generations may yet develop taste preferences for healthy, fresh food with minimal sugar, salt, and fat, especially if they start on such a template of foods. There is much work to be done in improving the standard American diet, however, and as long as a healthier diet is more expensive, those of limited means will remain dependent on cheaper, highly processed food products full of empty calories, what Sidney Mintz in another context called "proletarian hunger killers."[35] There will still be babies who receive substandard food until the struggling classes have a surer foothold in our economy, which today sees the greatest disparity between rich and poor since the Gilded Age. Globally, in the ensuing decades, as more people in developing countries crowd into cities, with minimal access to fresh fruits and vegetables, their diets will become more industrialized and similar to ours. They will experience the double burden of malnutrition, the health conditions that come not only from not enough food but also from the problems of the wrong kinds of food, the highly processed, energy-dense food of our culture.[36] Perhaps they can learn from our mistakes and avoid the worst excesses of our highly industrialized food system even as they benefit from its strengths.

It is interesting to consider Albala's and Laudan's diverging perspectives regarding the industrial food system in light of the story of commercial baby food. As Laudan explains, the historical trajectory of industrial processed food allowed the development of complex societies, provided for a more democratic approach to consumption, and gave people the soft, delicious and generally safe food supply they desired. Yet as Albala contends (and few would deny), the industrial food system also creates problems, including health issues resulting from too much highly palatable food as well as environmental degradation wrought by industrial production. Instead of throwing out the entire system, Laudan maintains, we can improve it: "Far from fleeing from them, we should be clamoring for more high-quality industrial foods." The story of commercial baby food suggests this is possible, as commercial baby food makers, responding to unfavorable nutrition studies and consumer distrust, created products acceptable to twenty-first-century notions of "good food."

Finally, the story of the invention and mass production of baby food in the United States magnifies the notion of convenience, a quality Americans value highly, as Albala and Laudan both note and assess differently. Baby food was so popular so quickly and consistently because it was time saving and convenient. It filled a need at a moment when the scientific discovery of vitamins dramatically changed the way adults approached infant feeding. Its little jars of products laden with sugar, salt, and starch were gateway foods to the industrialized American diet that blossomed in the mid-twentieth century. Using them made mothers feel confident and modern. Commercial baby food made it easier and more convenient for women with small children to enter the paid workforce and stay there.

Baby food was part of the twentieth-century food revolution, evolving in taste and quality but maintaining its character of convenience. But over the decades, in exchange for convenience, Americans were compelled to give up, or gave up willingly, other important qualities, to varying degrees at various times, including taste, nutrition, quality, commensality, and rules circumscribing when, where, and what to eat that helped control appetite and maintain healthy eating, qualities that Albala highlights and validates in his essay. Convenience is not just embodied by fast food, but is endemic to American life in all walks. Slowing down to cook a meal, to steam and mash a sweet potato for one's baby, helps recapture some of those lost qualities and values, but almost invariably at the expense of convenience. All are trade-offs, part and parcel of an expansive and powerful American food system and culture.

NOTES

1. See Amy Bentley, *Inventing Baby Food: Taste, Health, and the Industrialization of the American Diet* (Berkeley: University of California Press, 2014); portions of this chapter are taken directly from the introduction.

2. Herman Frederick Meyer, *Infant Foods and Feeding Practice: A Rapid Reference Text of Practical Infant Feeding for Physicians and Nutritionists* (Springfield, Ill.: C. C. Thomas, 1960), 143.

3. Alice L. Wood, "The History of Artificial Feeding of Infants," *Journal of the American Dietetic Association* 31, no. 5 (May 1955): 21–29, 24.

4. Judson Knight, "Gerber Products Company," in *Encyclopedia of Major Marketing Campaigns*, ed. Thomas Riggs (Farmington, Ill.: Gale Group, 2000), 664–67.

5. Joshua A. Boyce, Amal Assa'ad, A. Wesley Burks, Stacie M. Jones, Hugh A. Sampson, Robert A. Wood, Marshall Plaut, Susan F. Cooper, and Matthew J. Fenton, "Guidelines for the Diagnosis and Management of Food Allergy in the United States: Summary of the NIAID Sponsored Expert Panel and Report," *Journal of Allergy and Clinical Immunology* 126, no. S6 (December 2010): S1–S58.

6. Bright I. Nwaru, Maijaliisa Erkkola, Suvi Ahonen, Minna Kaila, Anna-Maija Haapala, Carina Kronberg-Kippila, Raili Salmelin, et al., "Age at the Introduction of Solid Foods during the First Year and Allergenic Sensitization at Age 5 Years," *Pediatrics* 125, no. 1 (2010): 50–59.

7. Jennifer J. Koplin, Nicholas J. Osborne, Melissa Wake, Pamela E. Martin, Lyle C. Gurrin, Marnie N. Robinson, Dean Tey, et al., "Early Introduction of Egg Might Protect against Egg Allergy," *Journal of Allergy and Clinical Immunology* 126, no. 4 (October 2010): 807–13.

8. Rachel L. Thompson, Lisa M. Miles, Joanne Lunn, Graham Devereux, Rebecca J. Dearman, Jessica Strid, and Judith L. Buttriss, "Peanut Sensitization and Allergy: Influence of Early Life Exposure to Peanuts," *British Journal of Nutrition* 103, no. 9 (May 2010): 1278–86.

9. George DuToit, Yitzhak Katz, Peter Sasieni, David Mesher, Soheila J. Maleki, Helen R. Fisher, Adam T. Fox, et al., "Early Consumption of Peanuts in Infancy Is Associated with a Low Prevalence of Peanut Allergy," *Journal of Allergy and Clinical Immunology* 122, no. 5 (2008): 984–91.

10. Perri Klass, "Advice Shifts on Feeding Baby," *New York Times*, 9 April 2013, http://well.blogs.nytimes.com/2013/04/08/advice-shifts-on-feeding-baby/, accessed 5 November 2018.

11. Tara Parker-Pope, "A Pediatrician's Advice on 'Green' Parenting," *New York Times*, 12 February 2010, http://well.blogs.nytimes.com/2010/02/12/a-pediatricians-advice-on-green-parenting/; see also Alan R. Greene, *Feeding Baby Green: The Earth-Friendly Program for Healthy, Safe Nutrition during Pregnancy, Childhood, and Beyond* (San Francisco: JosseyBass, 2009).

12. Elaine Louie, "Fireproofing Young Palates," *New York Times*, 16 August 1995, www.nytimes.com/1995/08/16/garden/fireproofing-young-palates.html; Gary K. Beauchamp and Julie A. Mennella, "Early Flavor Learning and Its Impact on Later Feeding Behavior," *Journal of Pediatric Gastroenterology and Nutrition* 48, no. S1 (March 2009): S25–S30.

13. Beech-Nut label, 1960s, Beech-Nut Plant, Canajoharie, N.Y.

14. Walter Sullivan, "Study Reports That Breastfeeding Is Declining," *New York Times*, 30 November 1967, 43.

15. Ronette R. Briefel, Kathleen Reidy, Vatsala Karwe, Linda Jankowski, and Kristy Hendricks, "Toddlers' Transition to Table Foods: Impact on Nutrient Intakes and Food Patterns," *Journal of the American Dietetic Association* 104, no. S1 (January 2004): S38–S44.

16. Kristen M. Hurley and Maureen M. Black, "Commercial Baby Food Consumption and Dietary Variety in a Statewide Sample of Infants Receiving Benefits from the Special Supplemental Nutrition Program for Women, Infants and Children," *Journal of the American Dietetic Association* 110, no. 10 (October 2010): 1537–41.

17. Hurley and Black, "Commercial Baby Food Consumption."

18. Mintel International Group, "Baby Food and Drink —US—June 2012: Reports," *Mintel Oxygen*, 1 June 2012, http://oxygen.mintel.com/display/590581/.

19. Stephen Daniells, "Beech-Nut CEO: Our Real Food Platform Is THE Place to Be with Millennial Moms," *FoodNavigator-USA.com*, 14 March 2014, www.foodnavigator-usa.com/Manufacturers/Beech-Nut-CEO-Our-real-food-platform-is-THE- place-to-be-with-Millennial-moms, accessed 20 March 2014.

20. Korky Vann, "A Passion for Mashing: Organic Baby Food Can Be Bought at the Store or Made at Home," *Hartford Courant*, 15 April 2010.

21. Jean Stevens, "Feeding Your Baby: For Some Women, Making Their Own Baby Food Makes Sense," *Herald News* (Passaic County, NJ), 5 September 2007.

22. Stevens, "Feeding Your Baby."

23. "A Fresh Generation of New Moms Reject a 107-Year Old Tradition: Jarred Baby Food," *Businesswire.com*, 6 May 2008, https://www.businesswire.com/news/home/20080506005600/en/Fresh-Generation-New-Moms-Reject-107-Year.

24. Amy Bentley, "Martha's Food: Whiteness of a Certain Kind," *American Studies* 42, no. 2 (Summer 2001): 5–29.

25. Katherine F. Michaelsen and Henrik Friis, "Complementary Feeding: A Global Perspective," *Nutrition* 14, no. 10 (October 1998): 763–66.

26. J. C. Drummond and Anne Wilbraham, *The Englishman's Food: A History of Five Centuries of the English Diet* (London: Pimlico, 1939), 68; Suzanne F. Adams, "Use of Vegetables in Infant Feeding through the Ages," *Journal of the American Dietetic Association* 35 (July 1959): 692–703.

27. Joan Jacobs Brumberg, *Fasting Girls: The Emergence of Anorexia Nervosa as a Modern Disease* (Cambridge, Mass.: Harvard University Press, 1988).

28. Roger Highfield, "Babies Fed on a Bland Diet 'Develop Taste for Junk Food,'" *The Telegraph*, 24 March 2005, www.telegraph.co.uk/news/uknews/1486309/babies- fed-on-a-bland-diet-develop-taste-for-junk-food.html.

29. Alan Greene, "2011 White Paper: Why White Rice Cereal for Babies Must Go" (2011), www.drgreene.com/ebooks/white_paper_white_rice_cereal.pdf, accessed 5 November 2018; Greene, *Feeding Baby Green*; Parker-Pope, "A Pediatrician's Advice."

30. Nancy F. Krebs, "Dietary Zinc and Iron Sources, Physical Growth and Cognitive Development of Breastfed Infants," *Journal of Nutrition* 130, no. S2 (February 2000): S358–S360; Gabrielle Palmer, *Complementary Feeding: Nutrition, Culture and Politics* (London: Pinter and Martin, 2011).

31. Susan J. McDonald, Philippa Middleton, Therese Dowswell, and Peter S. Morris, "Effect of Timing of Umbilical Cord Clamping of Term Infants on Maternal and Neonatal Outcomes," *Cochrane Database of Systematic Reviews* 7 (11 July 2013).

32. Z. Y. Ong and B. S. Muhlhausler, "Maternal 'Junk-Food' Feeding of Rat Dams Alters Food Choices and Development of the Mesolimbic Reward Pathway in the Offspring," *Federation of American Societies for Experimental Biology Journal* 25, no. 7 (2011): 2167–79; "'Junk Food' Moms Have 'Junk Food' Babies," *Science Daily.com*, 24 March 2011, https://www.sciencedaily.com/releases/2011/03/110323105200.htm.

33. Greene, *2011 White Paper.*

34. Gill Rapley and Tracey Murkett, *Baby-Led Weaning: Helping Your Baby to Love Good Food* (London: Vermilion, 2008).

35. Brenda Goodman, "Study: Healthy Eating Costs More," WebMD, 4 August 2011, https://www.webmd.com/diet/news/20110804/study-healthy-eating-costs-more#1; Sidney Mintz, "Quenching Homologous Thirsts," in *Anthropology, History, and American Indians: Essays in Honor of William Curtis Sturtevant*, ed. William L. Merrill and Ives Goddard (Washington, D.C.: Smithsonian Institution Press, 2002), 349–56.

36. Conversation with Katherine Kreis of the Global Alliance for Improved Nutrition (GAIN), 18 November 2013, notes in author's possession.

10

Home, Cooking

Why Gender Matters to Food Politics

TRACEY DEUTSCH

Many pieces in this collection speak to noisy debates, for instance about appropriate foods for infants or the social politics of nutrition. This essay, however, introduces a new question—at least one that is often overlooked. How should we address questions of gender, and particularly questions of the past and present nature of women's cooking, in contemporary food politics? With a few important exceptions, neither scholars nor activists nor food critics typically spend much energy debating what these calls for home cooking (or other food reforms) might mean to gendered divisions of labor, to questions of gendered equity, or to the lives of women. Often this is because people think the problem of women's domestic responsibilities has been solved, or that it is a selfish and naive issue in the current context, or that housework was unimportant to begin with (or, sometimes, all of the above). But talking about women's work isn't simply airing old laundry.

Thinking about gender gives us new questions to ask and new options for moving forward in food politics. Historical romanticization, as many in this collection have suggested, can too easily accompany calls for home cooking. But it's often a particular romanticization of work that *women* have done. If we can't think in gendered terms, we can't really analyze these narratives. Attending to the mundane tasks that made up "women's work," understanding how romanticized notions of the past have long been used to constrain women, and expanding our vision of the past to encompass the "many stories" that make up women's history of cooking is crucial. A smarter women's history offers ways to counter current narratives—a way to think beyond the constraints that so often make it hard to move forward in food politics. It helps us to connect what can seem like individual consumer choices to larger social systems that also need changing, a problem that Steve Striffler also identifies in his article (chapter 3). In fact, women's history has a lot to teach us about the past, and also about the present.

History looms large in much of the most important recent writing about food. Pastoral nostalgia celebrates free-range beef and chicken farms and the farm families who have retained or reinvented older forms of sustainable farming. Critics of commercially farmed or processed foods lament the loss of old agricultural skills and foodways and salvage "heirloom" seeds and breeds. Gary Nabhan, one of the first U.S. writers to describe his own local eating effort, rued the "ancient culinary melodies [that] are being drowned out by the noise of that transnational vending machine."[1] Most recently, the diets of "primitive" peoples have been marshaled to justify "Paleolithic" or "whole foods" diets.[2] In fact, appeals for a return to "real food" and "home cooking" are often the linchpin in plans to reform the food system. After all, the point in much alternative food work is not to simply *grow* different food, but to get people to *eat* different foods. Media personalities promise that home cooking will remake diets, the ways families spend their time, and entire food systems. Michael Pollan, perhaps the best-known critic of conventional food systems, makes this point succinctly in his recent book *Cooked.* He was surprised, he writes, to realize that the single most important thing he could do to improve his family's health and well-being, to make the food system healthier and more sustainable, to "achieve a greater degree of self-sufficiency," and to come to a deeper appreciation of nature was simply to cook.[3] Much recent writing shares this commitment to home cooking, including Ken Albala's essay for this collection.

As Pollan and many others acknowledge, for most of human history and much of its present, day-to-day family cooking has been the responsibility of women. Understanding the ways in which Pollan and other writers evoke gender and women's history is crucial to gaining leverage over their rhetoric of home cooking and shifting the debate about what and how we should eat and how the food system should change. Many of us agree that much has gone wrong in the contemporary food system. I share the assertions in many of this collection's essays that the contemporary food system needs attention. For instance, we would agree that the needs of small farmers and of the environment have been eclipsed by the demands of modern corporations. And we are alarmed that too many Americans eat unhealthy food. Like Charlotte Biltekoff and Steve Striffler, I am both sympathetic to the idea of equitable, healthy, and environmentally sustainable production and distribution—and also mindful of the ways similar calls for reform have served as vehicles for moralizing and for maintaining gender, racial, class, and other social hierarchies.

Acknowledging women's diverse histories of unpaid and paid work around food is crucial to achieving equity and to avoiding elitist moralizing. Paying attention to women's history draws our attention to crucial nodes in food systems and forces us to ask what is being "sustained" in sustainable foods. It demands that we acknowledge that household meals require labor (both paid and unpaid). Most importantly, women's history leads us to consider the well-being and the rights of the people who do that labor. It asks that we rethink the "home" in home cooking. In so doing, it helps us to think outside the box of contemporary food politics.

This chapter focuses on three themes. The first is the presence of gender and women's history in the home cooking discourse. Historical stories about women's food procurement and cooking in the past often are used to criticize contemporary food system (e.g., the move away from home cooking). They also are used as models for how we all ought to behave. For example, serving "traditional" food (especially "American" food), carefully prepared at home is held up as a model of what should be recaptured in the present. In other words, claims about women's history shape calls to take on responsibilities for home cooking in the present and are used to make such work sound pleasurable and fulfilling.

My second theme is how *long* this particular discourse has existed. For at least a century and a half, Americans have been told that American women have moved away from home cooking, that this has endangered families, and that they should return to older, better cooking ways of their grandmothers. They are asked to reinhabit the mystical women's home cooking traditions of yore. Since the 1850s, women's "failings"—indeed their "laziness"—has been blamed for problems in the food system. Nostalgic evocation of the foodways of older women, the celebration of older women's common sense, domesticity, and skill, has often gone along with criticizing the present generation of mothers for stubbornly insisting that they need to spend their time on something other than cooking. So when contemporary advocates of laborious home cooking respond to what they see as a new social problem, they are actually recapitulating an old complaint. In so doing, their solution repeats long-standing efforts to rein in women.

My third theme is that this effort to discipline people via nostalgia for women's past has meant highlighting *some* women's histories—and marginalizing the histories of others. It makes one unified narrative out of a much more complex past. Women who worked as cooks for other households, or on assembly lines of increasingly large food corporations, or in farm fields or migrant camps, are largely overlooked. The foods that women of color served and ate in their own families are often also overlooked and are too often cast aside as unhealthy.

Sometimes in this collection, and often outside it, the most powerful way in which race and class operate are when food writers overlook these entirely, obscuring the ways their memories are rooted in very particular experiences of race, class, and, often, privilege. The return to home cooking often accompanies paeans to "American" food, or to nurturing grandmothers, or to "traditional" food that "everyone" enjoys. And it is premised on a sense that "we" have moved away from this food. Women's history gives rise to the demand that we ask who the "we" is and what our past relationship with food might give us in the present.

Finally, I ask where this analysis leaves us. Attending to women's history does not mean rejecting calls for change. Instead, women's history pushes us toward a broader, intersectional vision of change. A fuller and more complicated women's history, one that departs from the single story told in contemporary discourse, *could* usefully expand the realm of food politics. While it is true that women did much of the labor of cooking for most of human history, it is also true that they did not always do so willingly. Women's histories reveal the charged nature of food work, the ways it has furthered both authority and oppression, and the uses that women's wages—as well as their food—have been put to in the past. It lets us see kitchens and home cooking as places of joy *and* power, authority *and* possibility, tradition *and* resistance. It even gives us some models for rethinking home cooking without turning to large corporations. Women's history reminds us of the hierarchies reinforced in households that need to be undone before food politics can truly be changed.

Much of contemporary food reform culture reinforces women's obligations toward meeting their families' food needs. For example, a host of advice columns from recent American newspapers and magazines (clearly directed at women) explain the how-to of shopping at farmers markets. Although they feature "family-friendly" advice, they also include instructions that would be difficult with children in tow—for example, to arrive early, carry your own bags, have conversations with farmers, and walk around the market all the way first before making any purchases.[4] Countless cookbooks also encourage these practices. A close look at a particularly successful, recently published cookbook reveals the entwining of gender, cooking, and broader changes in the food system. Lisa Leake, author of the *New York Times* bestseller *100 Days of Real Food,* describes her journey to remaking her family's eating as a journey away from her roots (in a home where her father did most of the cooking of "real" food and her mother "was more likely to fall prey to convenience foods"). "Real" food, she asserts repeatedly, is a way of life: eating this way requires checking with restaurants before agreeing to eat there, sending her husband off on business trips with snacks she has prepared, and elaborate entertaining to showcase one's efforts (e.g., hosting a large barbeque that featured grass-fed burgers and specially ordered buns). She is clear that this work stems from her commitment to family—the book is dedicated to her "amazing husband and daughters, who have wholeheartedly supported me throughout this journey" (the journey toward more cooking for them).[5]

This is a broader phenomenon. Allison Alkon's insightful fieldwork at Bay Area farmers markets revealed that most customers were women, that most farmers were men, and that women understood themselves as doing this extra shopping in service to their families. Similarly, sociologist Rebecca Som Castellano describes "the persistence of gender inequality in the division of labor in food provisioning" among households who pursue alternative food, as measured by women's ongoing responsibility for food provisioning and the amount of time that men versus women spent procuring household food. And most recently, Sarah Bowen, Joslyn Brenton, and Sinikka Elliott study how women of different races and classes negotiate this same pressure.[6]

Despite the questions these findings raise about a key task in local eating (namely, shopping), decisions about dividing up household tasks are typically simply not treated in food literature. Barbara Kingsolver, novelist and essayist, narrated her influential book *Animal, Vegetable, Miracle* through accounts of family-operated gardens and farms and the local eating practices of several households, including her own. Kingsolver openly acknowledges that did much of the work in her family's year of eating local, without much commenting on the negotiation of household tasks. Similarly, she lauded an Amish farm family in which the husband worked the fields and the wife, daughter, and daughter-in-law oversee household eating in apparent harmony.[7] Other accounts echo this silence. Craig Goodwin, a Presbyterian pastor, photographer, and avid gardener in Washington State who developed a book after blogging about his family's year of local eating and consuming, was rarely asked about how cooking worked in the household, although when he was, he was clear that his wife (and copastor) had done virtually all of it.[8] In these accounts, and in many others, women's work is crucial to the food that is consumed. But the ways in which women were pulled into that domestic arrangement go unremarked. Apparently, it is simply not a question that needs to be asked.

Many are careful to note that of course men, too, can and should cook. Indeed, this clearly happens in many homes.[9] Nonetheless, it is plain that if men don't cook, then women must. The stakes of refusing to cook homemade food are simply too high—we cannot sacrifice our very humanness, family coherence, or the health of ourselves, our partners, our children, and the planet—all things to which home cooking has been linked.[10] As Ken Albala asserts in this collection, "Not only is cooking central to what it means to be human, but . . . cooking from scratch using whole ingredients and gathering around the dinner table may itself be essential to our happiness." By this logic, resisting the call of kitchen work, for whatever reason, puts one's world at risk.

Perhaps the starkest example of this is Richard Wrangham's fascinating book *Catching Fire: How Cooking Made Us Human*. Wrangham, a Harvard biological anthropologist, argues that the act of cooking food was crucial to human evolution, allowing *Homo sapiens* to eat nutritionally dense foods, to eat and move more quickly, to develop the kind of social systems we currently have, and eventually to become "human." The sexual division of labor in which women were responsible for cooking was, according to Wrangham, "crucial to making us who we are" and also the result of patriarchy. It was unfair to women. But he does not offer any sense of what might be done to check patriarchy, given that it has resulted in cooking and that cooking is crucial to our humanity. The book concludes only with the admonition that we ought to cook more "real food."[11] Even among people who are "uncomfortable" with the politics of cooking, its results—our humanity, our social structure—mean that cooking cannot be rethought or avoided. It is simply too bad for women if they don't want to do it.

Wrangham's sense of women's long-standing attachment to cooking is important for other reasons also. It draws our attention to the historical argument, sometimes explicit and sometimes implicit, that accompanies calls for a return to home cooking. Put simply: if women were responsible for cooking, and cooking declined, then clearly it is because women made poor choices. For many authors, the wrong turn in American foodways accompanied women's new openness toward "convenience" foods and commercially prepared meals.

Mass retail, eating out, and particularly the explosion in processed foods are all explained as the neat results of women's (naive) preferences for their own pursuits, and, often, for paid employment. This is typically pegged as beginning with the 1960s and 1970s women's movements. Barbara Kingsolver argues that women, in particular, were strung along by the false promises of feminism. She explains: "When we traded homemaking for careers, we were implicitly promised economic independence and worldly influence. But a devil of a bargain it has turned out to be in terms of daily life. We gave up the aroma of warm bread rising, the measured pace of nurturing routines, the creative task of molding our families' tastes and zest for life; we received in exchange the minivan and the Lunchable."[12] For Kingsolver, female food shoppers were easy prey for marketers' messages and the promise of feminism.

Similarly, Mark Bittman, author and food writer for the *New York Times*, points to the effectiveness of large mass-marketing campaigns in encouraging the spread of processed foods. In his landmark book *How to Cook Everything*, Bittman writes, "Contemporary marketing has convinced many people that convenience food is not only quicker than home cooking, but also better and cheaper. In fact, it's worse."[13] Michael Pollan's 2009 article on the sad state of Americans' foodways takes a different tack, more explicitly naming women's responsibility for the shift. He explains: "It's generally assumed that the entrance of women into the workforce is responsible for the collapse of home cooking, but that turns out to be only part of the story. Yes, women with jobs outside the home spend less time cooking—but so do women without jobs. . . . Women with jobs have more money to pay corporations to do their cooking, yet all American women now allow corporations to cook for them when they can." In other words, it is women—as a group—who have allowed corporations to take over home life. Pollan repeats this more pithily in his more recent book, *Cooked*. "The outsourcing of much of the work of cooking to corporations has relieved women of what has traditionally been their exclusive responsibility . . . making it easier for them to work outside the home and have careers." It is not surprising that Pollan credits marketers' appropriation of a ready-made "rhetoric of kitchen oppression" with the end of real cooking.[14] It seems that for him, women, and particularly women's liberation movements and women's employment, are the problems.

While they may be unaware of it, these accounts by liberal and progressive food writers are not dissimilar to those forwarded by more conservative writers. Evangelical Christians committed to alternative foods share the belief that this division of labor is natural. For example, the founder of "Keeper of the Home" (a self-described mother of five, "drinker of fair-trade coffee with maple syrup and raw cream," and "Lover of Jesus") wrote excitedly of Kingsolver's book that it was rare to find a "feminist" who could discuss "the losses and burdens accompanied by the removal of wives and mothers from the home." Joel Salatin, a farmer made famous for his production of highly coveted meat and eggs on his sustainable, organic farm, is also staunchly committed to libertarian politics and evangelical Christianity. Salatin links his innovative farming and animal husbandry to a broader argument about limited government; a return to a household economy; "Christian," patriarchal family structures; and the benefits of entrepreneurial capitalism. His internships for would-be farmers have only recently been opened to women and are closed to non–U.S. citizens.[15]

The idea that American women's desire for more public authority and autonomy and for waged work brought about the decline of American food is widespread. Sociologist Ann Anagnost finds that evangelical Christians share with many advocates of food reform a resistance to women's move out of the home and a sense that feminism had much to do with this. Other scholars, too, have shown how dictates to cook and produce food at home are shared across lines of liberal and conservative communities. Both groups increasingly hold the belief that more food needs to be cooked at home and the question of who does that cooking is beside the point.[16] In this setting, asking about household labor inequities is akin to questioning the value of home cooking—and of civilization—itself.

Despite the urgent and heated claims made about women's turn away from home cooking in the late twentieth century, this is actually not a new "problem." Women have long been told that they've fallen away from their earlier, better, cooking methods. And they've been encouraged to return to cooking purer, more "American" food to raise healthier, responsible, cohesive families and citizens. Holding up earlier generations of women's foodways as a model for contemporary women is itself an American tradition.

Calls to reinvigorate household crafts and housework along these lines began at least 150 years ago, and probably earlier, but one example will suffice. In 1851 New England minister Horace Bushnell coined the phrase "the age of homespun" to refer to a bygone era when, he said, homes were sites of craft and skilled production and materials were natural. Bushnell's nominal reason for writing was the supplanting of homespun fabrics with industrially produced cloth, but his real concern was social change, and particularly what he saw as a new materialism and consumption. Bushnell valorized earlier

New Englanders' enjoyment of simple entertainments, epitomized by an evening of singing and the "home-grown exotics" with which the "good housewife" entertained: "doughnuts from the pantry, hickory nuts from the chamber, and the nicest, smoothest apples from the cellar." At the center of this world, for Bushnell, was women's dedication to household production and family life. Earlier marriages, for instance, evidenced a belief that "women are given by the Almighty not so much to help their husbands spend a living as to help them get one." This was a contrast to his own day, when, he said "many appear to assume" that "the woman is given to the man to enjoy his living" by going out and spending it.[17]

There is much that is striking about Bushnell's faith in women's domesticity, but I want here to draw on the three foods he mentions—the apples, nuts, and doughnuts. The celebration of women's serving of simple homemade "Yankee" food was revised and repeated in later years, and so too was the linking of this food to the stabilizing domesticity created by the selfless work of women of yore. For instance, in the late nineteenth century Ellen Swallow Richardson and a group of middle- and upper-class women started an effort to educate both upper-class women—who they worried had become dependent on commercial food—and the working-class women whom they expected to serve these upper-class families. In particular, they were concerned that both groups of women get used to "American" cooking. Dubbed "The New England Kitchen," the curriculum focused on what struck these native-born women as useful, everyday dishes (based on an invented New England past). This mode of cooking was enshrined in the Boston Cooking School's famous cookbook (later known as *The Fannie Farmer Cookbook*), but it was certainly not restricted to the East Coast. By the 1920s, cooking schools had sprung up in urban areas across the country to teach this version of healthy cuisine. Students—especially working-class people of all ages and young people—were encouraged to adopt fresh foods, to use simple, old-fashioned ingredients, and to cook for themselves rather than take advantage of canned goods, restaurants, delicatessens, and other sources of processed foods. Some writers linked the use of spicy food or purchased food to breakdowns in families or educational deficits.[18]

The 1950s and 1960s saw calls to return to old-fashioned, nationalist, household craft dissipate—but without disappearing. "If I were to design a coat of arms for our country," wrote the staff of General Foods in the voice of Betty Crocker, "a pie would be the main symbol . . . for pie is part of our history and tradition."[19] Indeed, a close reading of *The Betty Crocker Cookbook* reveals that for all the celebration of modern streamlined techniques, the authors embraced (and invented) American "traditions" (like pies and baked beans) with women's cooking at its heart. The first page for the section on breads quotes a poem: "An ancient rite, as old as life is old: A woman baking bread above a flame."[20] (Indeed, this message may have resonated strongly; for all that the 1950s and 1960s are associated with embrace of "convenience" and processed foods, historical scholarship shows that women resisted the messages of marketers of processed foods, modifying recipes and rejecting other suggestions out of hand.)[21] Increasingly, we can see the midcentury as a time that modified but reinscribed the discourse of women's home cooking of "real" American food. Homes created citizens, and the food served in them determined people's "fitness" as Americans.

Similar imperatives to identify an "American" cuisine, and similar advice about how to achieve it, circulate today. Barbara Kingsolver rhetorically, and disturbingly, asks: "Will North Americans ever have a food culture to call our own?," as if food reformers' goal is to construct a homogeneous and stable set of foods that erase differences of class, race, and history. Nina Planck's best-selling *Real Food: What to Eat and Why* offers several qualities of "real food" all of which center on links to the past. "Real food is *old,*" she offers—and then goes on, somewhat repetitively, to define real food as also "traditional." Her examples include "nice steak," "crispy . . . roast chicken," and mashed potatoes. In a turn of phrase that neatly erases cultural difference and universalizes a very particular cuisine, Planck assures her readers that "people everywhere love" these "traditional foods." Such appeals, in addition to homogenizing contemporary diets in North America, erase the dispossession of indigenous people and the oldest food traditions on the continent. Kingsolver and Planck might disagree about what counts as traditional food, but their commitment to it remains the same.[22]

Despite dramatic changes in the food distribution system, in work, families' lives, and notions of citizenship and belonging, this nationalist, Anglo-American advice to women about what food they should serve and why has not changed very much in 150 years. Even the same food shopping strategies have been proposed for over a century. Readers of the *Chicago Tribune* in the early twentieth century were regularly urged to purchase soup meat because it was so cheap, to make use of leftovers in meal planning, to travel to public markets and buy direct from farmers, to cook foods to their husbands' and children's liking, and to purchase fruit and vegetables only when in season (but, contradictorily, to serve beans and vegetables as the center of most meals). Speaking nearly one hundred years later, Rachael

Ray urged, among other things, that women buy in bulk, blanch and freeze vegetables that are on sale, shop at farmers markets, use coupons, and avoid unhealthy processed foods. Acknowledging financial pressures is certainly admirable, but Ray also resists acknowledging the complications of contemporary life: for example, that poorer families are less likely to have the large cooking equipment required for blanching and freezing, and that blanching and freezing vegetables might be two tasks too many for many households. Instead, like many food writers, Ray vividly evokes the past: "We need to go back to the way our grandparents prepared food. Instead of buying pieces of chicken, buy a whole chicken."[23] Past and present prescriptions thus closely resemble each other; women are being told to do the same basic tasks to accomplish the same goals of health and thrift—and that these are still their responsibility.

Here it is important to pause and take stock. The advice that women should emulate the past and the reminder that their homes were meant to be peaceful sites of citizen-creation has been consistently reiterated. This tells us something. It tells us especially that the advice has not been followed. If women have had to continually be told to do exactly the same thing, clearly the advice is not working.

The problem, I would contend, is with the idea that women can be helped with simplistic advice. Women have stubbornly refused to live within the dictates of the cuisines or the shopping advice others proposed. A fuller women's history offers a fuller sense of what has constituted "women's" traditions and fuller possibilities for rethinking cooking in the present. A fuller women's history means seeing that women are a diverse lot, as different as the food they cooked, the conditions under which they lived, and the goals they had for themselves and their meals.

Seeing the work of women who were household servants, for instance, clarifies that while families' mealtimes have often been joyful events, they have also shored up class and racial hierarchies. Through the early twentieth century, many upper-class families and those middle-class families who could afford it had their meals cooked by others—women of lower status and resources, often African American women and men. Rebecca Sharpless's work powerfully demonstrates the conditions of such work—the surveillance, the racism and classism, and the enormously long and difficult physical tasks involved. Scholars of Asian America have begun sketching out a similar story for Chinese American cooks and household servants.[24] Kyla Wazana Tompkins's brilliant reading of the politics of eating in nineteenth-century

American literature and culture demonstrates the centrality of food and cooking to the workings of America—not the democratic America of the nationalist food writer's imagination, but a racially unequal one.[25]

Food and cooking could also be a benefit for working-class women and women of color, although not for the reasons usually given; cooking in someone else's kitchen was a resource wielded powerfully by generations of African Americans—a source of pride, of skill, of wages, and a source of food itself, which could be "redistributed" from the kitchens of the privileged to those of the workers.[26] Meals were a site of resilience in the face of daily hardships, a means of resistance against systems that were designed to perpetuate inequality and that hardly cared for the sustenance of black children. That is to say, cooking was, and was broadly understood to be, effortful, laborious, and purposeful. Eating, provisioning, serving—these were highly political acts performed out of antiracist and familial commitments, as much as pleasure and love.

A fuller view of women's history also shows us the diverse sources of family food. The "crisis" of domestic servants that began during World War I meant that more and more American women cooked for themselves, although of course it was precisely then that processed foods became most available and, perhaps, most necessary. Working-class Americans had long taken advantage of food cooked elsewhere—saloons, delicatessens, bakeries, peddler carts—and now used processed foods like canned goods. These trends increased over time.[27] Cooking like our grandmothers, or great-grandmothers, would mean understanding the place of processed foods in their strategies.

Processed foods did not alleviate the work of cooking. In the mid-twentieth century, the new canned and frozen goods promised convenience and wider variety in Americans' diets, but they also required new ways of cooking (re-creating familiar foods with them, for instance).[28] Shopping in larger stores farther from home required cash with which to make large purchases, much more extensive meal planning than in the past, and more laborious trips to the store. In 1974, the sociologist Joanne Vanek estimated that although the time spent cooking had dropped considerably in recent decades, it was nearly canceled out by the time women spent planning, driving, and shopping.[29] Writing sixteen years later, the feminist sociologist Marjorie Devault found very similar efforts made around meal planning, price comparison, and travel to and from the store. Of course, food shopping posed special frustrations for the poor and increasingly the urban poor. Before, during, and after the riots and insurrections of the mid- and late 1960s, large mass retailers forsook urban centers for the fringes of cities and suburbs. It was nearly a truism by the late 1960s that the poor "paid more" for daily necessities—and often found goods of lower quality in stores near their homes.[30] Food work, wrote Marjorie DeVault, reveals "the complex ways that women are themselves drawn into participation in prevailing relations of inequality" even as they also perform vital and rewarding tasks.[31]

As this feminist scholarship suggests, procuring and cooking food was *never* easy. Taken as a whole, these histories show us that processed food and wage labor has long been at the center of the meals celebrated as "home cooking."

Perhaps because of the long history of negotiating the work of cooking, women's history also left a legacy of inventive ways of reorganizing food and cooking—beyond the "tradition" of private homemade meals prepared by mothers and full of pure ingredients. The same moment that gave rise to the Boston Cooking School and the New England Kitchen also gave rise to public feeding programs and efforts to lighten the work of individuals' caring for families. Generations of progressive women activists pioneered public meal programs. However limited their cuisine choices, they point us to the widespread realization that food could be distributed publicly as well as privately—and to good effect.

It isn't surprising that progressive women pioneered in this. Many lived outside nuclear families in settlement houses, as single adults, or as partners of other women. They saw that the working-class people around them knew as much about cooking and eating as they did but lacked the resources to do it. They understood, sometimes because they themselves experienced it, that meals did not need to be produced by mothers, or eaten in a family setting, to be nourishing or productive. Many of them offered critiques of the dictates of middle-class family life. Over time, the cooking classes they taught and the food they encouraged reflected the cuisines they were surrounded by, challenging notions of "American" food.[32]

And so they created a litany of ways of reimagining meals. Today's school lunch programs remain a legacy of their work. Other of their ideas have been obscured: faced with an influenza outbreak after World War I and increasing frustration about the time required to cook, women formed "community kitchens" to prepare meals for whole neighborhoods of subscribers. Some of these existed for decades. Some feminist activists, particularly socialists, imagined other ways of rearranging space—housewives' cooperatives, kitchenless houses, and community dining halls for instance.[33] Implicit in these models was that households could survive even if meals were eaten outside of family settings. Indeed, sometimes they thrived. Food could sustain new social formations, not only conventional families and not only within the walls of a family home.

Nothing makes this point more clearly than second-wave feminism—the movement so suspect in contemporary food politics. Advocates of "women's liberation" certainly railed against the frustrations of home cooking, framing daily cooking as one more site of women's domestic isolation and oppression. Pat Mainardi included "buying groceries, carting them home and putting them away; cooking meals and washing dishes and pots" on her list of "dirty chores" in the widely circulated 1970 essay "The Politics of Housework."[34] And in 1975 Martha Rosler, a pathbreaking performance artist, filmed herself using kitchen implements to produce violence, and occasionally music, in one of her first pieces, "Semiotics of the Kitchen."[35] Rosler, like so many of her peers, criticized the drudgery, the mundane mindlessness, that so many expected of women and their home cooking.

Women reimagined cooking as part of reimagining the kind of world they wanted to inhabit. Fixing one thing (say, nutrition) required fixing others (say, household relations). Ideas ranged from demands that men join in mundane housework, including daily cooking chores (moderately successful), to celebrations of cooking as a traditional women's activity that needed reclaiming (as in *Laurel's Kitchen*) to efforts at "communal" kitchens and cooperative living that questioned nuclear families.[36] For instance, Pat Mainardi's essay went on to detail her strategies to get her husband to help with the housework. Alix Kate Shulman published her "Marriage Agreement" with her husband. Perhaps most strikingly, one group of socialist feminist women of color named their publishing house "Kitchen Table Press"—an indication of how seriously they took these spaces, not simply as sites of family or order, but as sites of energizing conversation, of possibility, of resilience and resistance. In other words, the charged spaces of kitchens have been used to rethink power relations more broadly.

In important ways second-wave feminist thinking predicts, rather than contradicts, some of the demands of today's food writers. The proposition that kitchen work should be equitable dovetails with the newer insistence that cooking can be a route to increased individual and social well-being. The difference is in how cooking is framed. Michael Pollan, and many others, argue that men and children ought to cook simply because cooking is such a generative, wonderful activity over and above any concerns with equity.[37] Earlier activists instead encouraged new ways of cooking precisely because it was not always wonderful. It was, instead, important and vital labor that needed to be shared if everyone were to be fully nourished.

Women's history allows us to include the well-being of the people doing the cooking as part of what should be fixed when we fix food systems. It allows us to look beyond the boundaries of "home" when we want to understand what happens within it. Looking at women's history reinforces the importance of cooking—but as something that connects people to other systems, including patriarchy. It makes homes more important and more complicated than the phrase "home cooking" might suggest.

Food preparation and shopping is still hard work today. Impoverished families continue to struggle to make inadequate wages and/or public benefits somehow work until the end of the month and to find food that household members and the folks who join a household for meals will eat and that they have the resources to prepare. For women of all classes and backgrounds, food remains the vehicle in which many, often contradictory, goals are to be accomplished—familial coherence and financial solvency, religious dictates and political commitments, class ambitions and household preferences. Prescriptive literature around shopping continues to urge women to do almost exactly what they were told to do a century ago—to buy in large quantities (but not too large, since that would be wasteful), to cook their own foods rather than buying processed foods, to prepare meals that fit current standards of nutrition and that also are "tasty," and to purchase cheaper cuts of meat and seasonally available foods. New innovations—coupons and frequent-purchaser cards—join long-standing dictates that meals needed to succeed on many levels: financial, nutritional, social, aesthetic, and cultural. Even the most locally grown of vegetables do nothing to eliminate the work of obtaining, cleaning, and serving, them.[38]

It has proven difficult to connect the promises of contemporary foodie-ism with calls for other kinds of justice. Too often, concerns begin and end with cooking and its pleasures, assuming domestic relations (and much else) will take care of themselves once that cooking's joys are prioritized. I argue here that these beliefs reflect a romanticization of the home and women's labor in it and a rejection feminist calls to value women's well-being and to question the division of labor within homes.

One example encompasses this remarkable erasure of questions of women's unpaid labor. In 1990, Barbara Kingsolver's short story "Homeland" appeared in an anthology. (The story had been published earlier in *Mademoiselle* magazine.) In the story, Lydia and Whitman relocate from a busy urban life to a rural cabin. Kingsolver tracks the near-breakdown of the couple's relationship, marked by Whitman's professional and personal frustrations, his retreat from domestic chores, and Lydia's taking these on. "I didn't apply for the job I'm doing here," Lydia says in the climactic scene. "I don't know how I got into it, but I know how to get out," she threatens as she storms off to her paying job.[39] As the couple tries to find new languages and new strategies for making their life bearable, Lydia misses the "uncomplicated affection" she has with the family dog. Compare this to the narrative of *Animal, Vegetable, Miracle,* published twenty-five years later. Here the issues of who does the housework and the dangers of rural isolation are barely raised.

Shifts in the intervening quarter century explain the changed narratives. By 2007, the food movement's emphasis on the benefits of home cooking eclipsed earlier concerns with the power relations that created the food. The tensions of domestic life, the intersection of racial, class, and sexual privilege that circle around meals—that is, the very forces that have made food such a compelling presence and identity marker in our lives—have been obscured by calls to "return" to home cooking. Our historical accounts of how we got to this moment and, consequently, what ought to happen to create a more just food system are murky, often blaming women for larger structural shifts or collapsing their many different stories into one narrative of how "women" used to cook.

My point here is not simply that it is too easy to overlook women's labor when we talk about food, nor that many of the prescriptions for change depend on conservative ideas of gender relations, nor that the terrible working conditions that produce food are too often eclipsed by claims of what food can do—although all of these statements are true. And my point is certainly not that we ought to unthinkingly reproduce ideas from the past; like Rachel Laudan, I think the past is no pattern book. Rather, I argue that we ought to use the past to rethink not just the kinds of food but the kinds of justice and equality we want in the future. Changing the food system will require—has always required—changing relations among those who live in households.

1. Gary Nabhan, *Coming Home to Eat: The Pleasures and Politics of Local Foods* (New York: W. W. Norton, 2002), 13–14.

2. See, for instance, the *Wise Traditions* journal of the Weston A. Price Foundation; Loren Courdain, *The Paleo Diet: Lose Weight and Get Healthy by Eating the Foods You Were Designed to Eat* (New York: Wiley, 2010); Richard Wrangham, *Catching Fire: How Cooking Made Us Human* (New York: Basic Books, 2009).

3. Michael Pollan, *Cooked: A Natural History of Transformation* (New York: Penguin, 2013), 1–2.

4. Molly Watson, "10 Farmers Market Shopping Tips," Spruce Eats, updated 26 August 2018, https://www.thespruceeats.com/farmers-market-shopping-tips-4067698/, accessed 15 March 2019; Sarah W. Caron, "Tips for Shopping the Farmers' Market," 21 June 2010, http://www.tablespoon.com/recipe-blog/2010/06/21/tips-for-shopping-the-farmers-market/, accessed 10 July 2010 (page discontinued); Lisa Borden, "Shopping Smart at the Farmers Market," http://crazysexylife.com/2010/shopping-smart-at-the-farmers-market/, accessed 10 July 2010 (page discontinued); Masterclass, "How to Shop at a Farmers Market with Alice Waters," https://www.masterclass.com/articles/how-to-shop-at-the-farmers-market-with-alice-waters#alices-guide-to-shopping-at-the-farmers-market/, accessed 15 March 2019; "How to Shop the Farmers Market for Dinner," https://www.loveandlemons.com/farmers-market-dinner/, accessed 15 March 2019.

5. Lisa Leake, *100 Days of Real Food: How We Did It, What We Learned, and 100 Easy, Wholesome Recipes Your Family Will Love* (New York: William Morrow Cookbooks, 2014), 1, 105–6, 4, 104.

6. Rebecca Som Castellano, "Alternative Food Networks and Food Provisioning as a Gendered Act," *Agriculture and Human Values* 32, no. 3 (2015): 461–74. https://doi.org/10.1007/s10460-014-9562-y; Sarah Bowen, Joslyn Brenton, and Sinikka Elliott, *Pressure Cooker: Why Home Cooking Won't Solve Our Problems and What We Can Do about It* (New York: Oxford University Press, 2019).

7. Barbara Kingsolver, *Animal, Vegetable, Miracle: A Year of Food Life* (New York: HarperCollins, 2007), 59–69.

8. Goodwin's blog was at www.yearofplenty.or, accessed 11 July 2010; his experiment received significant press coverage, including J. Fowler, "Interview with Craig Goodwin, Author of Year of Plenty," http://sustainabletraditions.com/2011/09/interview-with-craig-goodwin-author-of-year-of-plenty/, accessed 15 March 2019; William Yardley, "Pastors in Northwest Find Focus on Green," *New York Times*, 16 January 2010, p. A11; http://www.fastrecipes.com/interesting-cooking-resources/year-of-plenty:-a-recipe-for-living-and-cooking-local-2010061240075/, accessed 11 July 2010. Cooking is described in more general, shared terms in Godwin's book. See Craig Goodwin, *Year of Plenty: One Suburban Family, Four Rules, and 365 Days of Homegrown Adventure in Pursuit of Christian Living* (Minneapolis: Sparkhouse, 2011). For an alternative to this silence on the question of division of labor, see Alisa Smith and J. B. MacKinnon, *Plenty: One Man, One Woman and a Raucous Year of Eating Locally* (New York: Harmony Books, 2007). Perhaps not coincidentally, James Mackinnon was the primary household cook.

9. Interestingly, there have been important studies of this movement, particularly among evangelical Christians. See Travis Warren Cooper, "'Cooking with Gordon': Food, Health, and the Elasticity of Evangelical Gender Roles (and Belt Sizes) on The 700 Club,"

Religion and Gender 3, no. 1 (January 2013): 108–24; Ann Anagnost, "Securing the Home Front: The Pursuit of 'Natural Living' among Evangelical Christian Homemakers," *Social Politics* 20, no. 2 (2013): 274–95.

10. Pollan, "Out of the Kitchen and onto the Couch," Mark Bittman promises that "conscious eating" will "help you lose weight, reduce your risk of many longterm or chronic diseases, save you real money *and* help stop global warming" and in addition "will be easier and more pleasant than any diet you've ever tried . . . [will] take less time and effort than your exercise routine, and [will] require no sacrifice" (Bittman, *Food Matters,* 1). For other examples of this widely circulating discourse see Alice Waters, *The Art of Simple Food* (New York: Clarkson Potter, 2007); Jamie Oliver, *Jamie's Food Revolution: Rediscover How to Cook Simple, Delicious, Affordable Meals* (New York: Hyperion, 2009). So pervasive is the discourse of family meals that it is used to frame even books making more complicated arguments. See, for instance, Janet Flammang, *The Taste for Civilization: Food, Politics, and Civil Society* (Urbana: University of Illinois Press, 2009). Flammang argues that meals' links to women's reproductive labors has led them to be devalued in modern capitalist structures—but her book is framed by the publisher as promoting household meals as key to civilization.

11. Wrangham, *Catching Fire,* 137, 206–7.

12. Barbara Kingsolver with Steven Hopp and Camille Kingsolver, *Animal, Vegetable, Miracle: A Year of Food* Life (New York: HarperPerennial, 2008), 126–27.

13. Mark Bittman, *How to Cook Everything,* rev. ed. (New York: Wiley, 2008); Bittman, *Food Matters,* chap. 6 ("Selling the Bounty"), 31–36.

14. Michael Pollan, "Out of the Kitchen, onto the Couch," *New York Times Magazine,* 2 August 2009, http://www.nytimes.com/2009/08/02/magazine/02cooking-t.html; Pollan, *Cooked,* 7.

15. "About," Keeper of the Home, https://www.keeperofthehome.org/meet-stephanie, accessed 20 June 2017, and "The Aroma of Warm Bread Rising and Other Things We'd Miss Out On," Keeper of the Home, 14 April 2009, https://www.keeperofthehome.org/the-aroma-of-warm-bread-rising-and-other-things-wed-miss-out-on. For Salatin's antagonism to state regulation and faith in supply and demand, see *Everything I Want to Do Is Illegal: War Stories from the Local Food Front* (White River Junction, Vt.: Polyface, 2007). Joel Salatin, "Re-birth of the Family Farm and the Household Economy" (2008), http://farmersandfreeholders.org/joel-salatin-re-birth-of-the-family-farm-and-the-household-economy, accessed 15 March 2019 (page discontinued). On the policy toward women on Polyface Farms, see "Is Back to Nature Farming Only for Men? The Two Faces of Polyface Farm," http://www.irregulartimes.com/polyface.html, accessed 5 July 2010 (site discontinued); A. Breeze Harper, "White Fragility and Joel Salatin's 'Good Food Framework,'" http://sistahvegan.com/2015/09/04/white-fragility-and-joel-salatin-daring-to-critique-the-mainstream-food-sustainability-movements-white-hero/, accessed 15 March 2019; Alison Hope Alkon, *Black, White, and Green: Farmers Markets, Race, and the Green Economy* (Athens: University of Georgia Press, 2012), 101. Salatin blames government bureaucracy for refusing to open the internship to non–U.S. citizens: http://www.polyfacefarms.com/apprenticeship/accessed, 15 March 2019.

16. On evangelical cooking, see Anagnost, "Securing the Home Front." See also Alkon, *Black, White, and Green,* 98.

17. Horace Bushnell, "The Age of Homespun," in *Work and Play* (New York: C. Scribner's Sons, 1881), 389.

18. Harvey Levenstein, *Revolution at the Table: The Transformation of the American Diet* (Berkeley: University of California Press, 2003), 5, 104.

19. General Foods, *Betty Crocker's Picture Cook Book* (Minneapolis: General Mills; New York: McGraw Hill, 1950), 293.

20. General Foods, *Betty Crocker's Picture Cook Book,* 83.

21. Jessamynn Neuhaus, *Manly Meals and Mom's Home Cooking: Cookbooks and Gender in Modern America* (Baltimore: Johns Hopkins University Press, 2003); Laura Shapiro, *Something from the Oven: Reinventing Dinner in 1950s America* (New York: Penguin Books, 2005).

22. Kingsolver, *Animal, Vegetable, Miracle,* 17; Nina Planck, *Real Food: What to Eat and Why* (New York: Bloomsbury, 2006), 1–3.

23. Jane LeBaron Goodwin, "Domestic Science," *Chicago Tribune,* 10 November 1905, 9; "How to Reduce the Cost of Living," *Chicago Tribune,* 22 December 1912, F6; Mary Eleanor O'Donnell, "Fighting Cost of Living," *Chicago Tribune,* 20 May 1913, 12; "Baskets on Arm; Wives to Market," *Chicago Tribune,* 12 January 1915, 1; Rachael Ray interview with Ali Veshi, CNN, recorded 10 March 2009, http://www.cnn.com/2009/SHOWBIZ/TV/03/10/lkl.rachael.ray/index.html, accessed 2 July 2010.

24. Rebecca Sharpless, *Cooking in Other Women's Kitchens: Domestic Workers in the South, 1868–1960* (Chapel Hill: University of North Carolina Press, 2010). See, for instance, Yong Chen, *Chop Suey USA: The Story of Chinese Food in America* (New York: Columbia University Press, 2014).

25. Kyla Wazana Tompkins, *Racial Indigestions: Eating Bodies in the Nineteenth Century* (New York: New York University Press, 2012).

26. See especially Psyche Williams Forson, *Building Houses out of Chicken Legs: Black Women, Food, and Power* (Chapel Hill: University of North Carolina Press, 2006).

27. Katherine Turner, *How the Other Half Ate: A History of Working-Class Meals at the Turn of the Century* (Berkeley: University of California Press, 2014).

28. Shapiro, *Something from the Oven.*

29. Joanne Vanek, "Time Spent in Housework," *Scientific American* 231 (November 1974): 116–20.

30. The definitive text on higher prices in poor urban neighborhoods in this era remains David Caplovitz, *The Poor Pay More: Consumer Practices of Low-Income Families* (New York: Free Press of Glencoe, 1963). See also Felicia Kornbluh, *The Battle for Welfare Rights: Politics and Poverty in Modern America* (Philadelphia: University of Pennsylvania Press, 2007), 116–19; Marjorie DeVault, *Feeding the Family: The Social Organization of Caring as Gender Work* (Chicago: University of Chicago Press, 1990), 67, 170–99.

31. Devault, *Feeding the Family,* 11.

32. On growing acceptance of "ethnic" foods, see Donna Gabaccia, *We Are What We Eat: Ethnic Food and the Making of Americans* (Cambridge, Mass.: Harvard University Press, 1998).

33. Turner, *How the Other Half Ate,* 128–29; Dolores Hayden, *The Grand Domestic Revolution: A History of Feminist Designs for American Homes, Neighborhoods, and Cities* (Cambridge, Mass.: MIT Press, 1981); Evanston Community Kitchen, https://evanstoncommunitykitchen.wordpress.com/about/, accessed 30 June 2015, and https://evanstonnow.com/event/education/bill-smith/2013-03-04/55018/womens-co-op-the-community-kitchen, accessed 15 March 2019.

34. Pat Mainardi, "The Politics of Housework," *Redstockings* (1970), https://caringlabor.wordpress.com/2010/09/11/pat-mainardi-the-politics-of-housework/, accessed 23 February 2019.

35. Martha Rosler, *Semiotics of the Kitchen* (1975), video (black-and-white, sound) MoMA Number: 718.19811975, 1975.

36. Stacy Williams, "A Feminist Guide to Cooking," *Contexts* 13 (Summer 2014): 59–61; Stacy Williams, "Personal Prefigurative Politics: Cooking Up an Ideal Society in the Woman's Temperance and Woman's Suffrage Movements, 1870–1920," *Sociological Quarterly* 58 (January 2017): 72–90.

37. Pollan, *Cooked,* 10–11.

38. Useful recent investigations of this problem are Tracie MacMillan, *The American Way of Eating: Undercover at Walmart, Applebee's, Farm Fields, and the Dinner Table* (New York: Scribner's, 2012); and Sarah Bowen, Joslyn Brenton, and Sinikka Elliott, *Pressure Cooker: Why Home Cooking Won't Solve Our Problems and What We Can Do about It* (New York: Oxford University Press, 2019).

39. Barbara Kingsolver, "Blueprints," in *Homeland and Other Stories* (New York: Harper & Row, 1990), 39.

Section V
Cooking and Eating Food

Section V reflects on perhaps the most subtle debate in this volume, and that is the importance of food and drink beyond the mere act of sustenance. Does thinking about food, cooking food, and eating skillfully or lovingly prepared food make us happier people? Clearly we need to eat and drink to survive, but should we just eat to live, or should food be more to us than necessity? Indeed, should we live to eat? This debate is significant because eating is something that all humans do and spend a lot of time doing. Should we therefore not also consume consciously, an act that might mean simply savoring the fragrance of a perfectly ripe peach, or enjoying the taste of an ice-cold Coke on a hot summer day? And isn't it important to eat with attention and intention? After all, as the chapters in this book show, eating and drinking always have both a meaning and a social impact.

In fact, one of the oddest aspects of American culture is our general dismissal of commensality. Most human cultures have considered food preparation and consumption, especially consuming food together, as essential to family, tribal, religious, and other social bonds. Some people would go even further and say that as social creatures, eating together makes us more socially adept and indeed happier human beings. However, in our highly individualistic society the value of eating and drinking together is probably honored more in the breach than in the observance. Perhaps, then, the best way to answer the question of whether food and drink is important for human happiness is for more people to take this question seriously in the first place. After all, we must all eat and drink to live, so why not eat and drink well? And while there is no agreed-upon definition of what constitutes eating and drinking well, surely reflecting on what that might mean must be part of a good life.

11

Is Thinking Critically about Food Essential to a Good Life?

ROBERT T. VALGENTI

It was a philosopher, Ludwig Feuerbach, who famously stated in 1862, "You are what you eat."[1] But rather than serve as a powerful capstone to an over-two-thousand-year tradition of thinking about the connection between food and humans as *the* philosophical issue—what does it mean to *exist*, to *be*, a human?—Feuerbach's statement functions as a curious, if not ironic, footnote in a long history of food's absence from the philosophical menu. Is food worthy of philosophical consideration? A glance back over the history of philosophy tells us that it has not been kind to the idea, and more importantly, that the issues of human alimentation have represented a problem for philosophy and a life dedicated to thought.

What is at stake in a "philosophy of food"? Such thinking would concern itself with all manner of food and drink, along with its preparation, consumption, and appreciation. One might consider such issues under the rubric of "gastronomy," which is defined by Brillat-Savarin as "a scientific definition of all that relates to man as a feeding animal," the goal of which "is to watch over the preservation of man by means of the best possible food."[2] But the science of gastronomy is not the dedicated task of *thinking about* gastronomy as one among the many possible ways that humans can or should exist in their world. In order to find a bit of focus for this potentially intractable idea, I propose a specific question that, in my view, captures much of the traditional disdain for this topic, as well as the possibility for its redemption: Is the philosophical consideration of gastronomy—a "philosophy of food" in the fullest sense—a necessary component of "the good life"?

For the purposes of this argument, I kindly ask the reader to accept two presuppositions. The first is organizational and will hopefully reduce the philosophical jargon employed in this essay: namely, that the arguments for and against a philosophy of food conform to the themes of rationality, universality, and purity. The second presupposition is philosophical: namely, that "the good life" is one that necessarily involves the practice of *critical thinking*, and, more specifically, critical thinking about what it means to be a human being and to live with other human beings. This presupposition is shared by all philosophers (but certainly not limited to them), even though they may differ on exactly what this involves and whether the good life is one thing or many things. Guided by these presuppositions, this essay will examine why a philosophy of food has traditionally been excluded from considerations of the good life and how philosophers past and present make the case for its inclusion.

GLANCING OVER THE MENU

The majority of philosophers have not taken much interest in food as a matter for philosophical consideration. There are those who, like dedicated dieters, politely avoid it, or push it to the side of the plate. Others, more prone to asceticism, condemn it as a threat to the higher and more abstract goals of philosophical reasoning. As a result, the philosophical menu—twenty-five hundred years strong and countless patrons served—has made an issue of food primarily through its conspicuous absence. When food has made it onto the menu, it is often considered a "special"—interesting, a change of pace, maybe even a bit adventurous. Most philosophers, however, are creatures of habit and, rather than take a risk, stick to the tried-and-true favorites. The sheer lack of published philosophy dealing with food and food-related issues gives some indication, to those in the field, of the general distaste of philosophies of food. And when food is a matter for discussion in published philosophy, it is usually ignored, downgraded, or dismissed with polite laughter.

Take for example Peter Singer's *Animal Liberation* (1975), one of the most important works of philosophy in the past forty years. While it is often heralded as an essential text for modern utilitarian ethics, and even environmental philosophy, it is rarely discussed as a "philosophy of food." Contemporary ethics has certainly been one of the areas of philosophy where food comes up most often. There is no doubt that food gives rise to ethical questions (What should we eat or avoid eating? Is this food good for the environment? Were the producers of this food paid a fair wage?) Yet these important questions often appear in a downgraded form as an "applied" philosophy of food—philosophical ideas and theories, developed in some other more purely philosophical context, are applied to issues that arise in the realm of gastronomy.[3] And even when philosophers develop philosophies of food in the manner that aims to critique the very presuppositions of philosophy itself, they are often framed in a way that is (intentionally or not) dismissive. One such technique is the play on words. Titles of books (no doubt suggested by well-meaning editors) such as *Food for Thought* or *Eat, Think and Be Merry* suggest that their subject matter is less serious than canonical works such as *The Critique of Pure Reason* or *Being and Nothingness*. Of course, there are more direct snubs. Who among us (we philosophers of food) hasn't heard from a colleague: "Oh, that's interesting; but what does that have to do with *real* philosophy?" And of course the fact that much of the outstanding recent philosophy of food is published in academic journals *not* dedicated solely to philosophy suggests (albeit indirectly) that the most prestigious philosophical journals have yet to acquire a taste for this sort of thinking.

The reasons behind the conspicuous absence of food from the philosophical menu, however, have much deeper roots than the ones outlined above. A defense of the "philosophy of food" must engage the very necessity of its existence and thus expose the inherent problem with a tradition that has all but avoided the issue. If we entertain the presupposition I suggested above regarding the good life and its inherently philosophical character, the table is set for an examination of the history of philosophy through the lens of some its prevalent conceptual habits: the belief that reason alone is the key to truth; the assumption that philosophical truths are universal; and the quest for a certain level of purity (whether it is in the definition of a thing's essence, or in the singular nature of the philosopher's quest for wisdom). And despite this somewhat arbitrary organizational schema, my intent is merely to highlight different themes that have undermined the acceptance of the philosophy of food and obscured its necessary position within what Socrates described as the "examined life."

Here I can only provide a summary overview of some of the exceptional research that, in recent decades, has addressed the gastronomical absences in the philosophical tradition. I will therefore stay within the limits of my proposed thesis and begin this examination by detailing the ways that some of the first philosophers—specifically Plato and Aristotle—portrayed the good life as one opposed to the sustained consideration and reflection upon food. I will then consider the problems associated with "taste" as a physiological sense of discernment and as a metaphor for cultural knowledge about the beautiful. From there I will highlight moments in the history of philosophy when food is taken seriously; and then I will conclude with an overview of the many recent arguments in favor of a philosophy of food.

A TRADITIONAL RECIPE FOR THE GOOD LIFE

Many of the arguments against the philosophical reflection on food can be found deep within the history of Western philosophy. In fact, it could certainly be argued that once philosophical thinking set off down a certain path—one that separated the rational, thinking aspect of human existence from its embodied, physiological aspect—the exclusion of food from philosophical consideration was as much a cognitive habit as it was a willful act of exclusion. But the origins of its dismissal are rather clear and rest firmly on the idea that the good life was a question to be shaped and answered primarily by a human mind that was distinct from a physiological body—one capable of thinking rationally, discovering universal truths, and distinguishing the pure essence of reality from its many particulars and distractions.

Plato's dialogues provide the most comprehensive picture we have of his philosophical teacher Socrates, but also provide a rich literary context in which food, drink, markets, and banquets abound.[4] Plato did not set forth a specific doctrine of ethics or a theory of "the good life" per se, but one can gather through his depictions of Socrates and the general pursuit of "the Good" that the philosophical topics and practices illustrated throughout the dialogues ultimately concern human flourishing (*eudaimonia*) and how such a state can be attained. The good life is one turned toward the rational soul and its improvement, a process most famously articulated in Plato's "allegory of the cave" as a movement away from the particulars of the physical world and toward the universal and pure eternal forms available to rational contemplation. Plato represents a tradition in Western philosophy that looks suspiciously upon the vicissitudes of physical experiences and human appetites. Food, and everything sorrounding its consumption and pleasure, fails to reflect the values of rationality, universality, and purity. One need only look at Plato's depiction of cooks and the role they will play in his utopian *Republic*. Barely artisans, their job is only to service the most basic needs of alimentation. Philosophers have no concern for such worldly pursuits, and when Socrates is asked in *Phaedo* about the importance of the pleasures of food and drink for philosophers, Socrates explains that philosophers "must be conerned with neither food, drink, nor sex. . . . The true philosopher despises them."[5]

Plato's arguments for the ultimate contemplation of "the Good" sets the stage for a hierarchical ordering of values that are mapped onto the human body and that persist long into philosophy's future. In *Making Sense of Taste* (1999), Carolyn Korsmeyer argues that Plato's negative bias toward the particulars of bodily experience leads to a value-laden "hierarchy of the senses" that shapes the entire course of Western thought and morality. Rational thought and the life of contemplation are often associated with sight and hearing, "higher senses" that allow for a more "objective" relation at a distance to the objects of this world. The "lower senses" of taste, smell, and touch require direct contect with the perceived object and are more "subjective." The former is thus akin to knowledge, while the latter to mere opinion. In the dialogue *Timaeus*, for example, the lower senses are associated with the destructive appetites, and the stomach in particular is the seat of the appetitive soul "that must be kept chained like a wild animal lest it overtake the whole being."[6] As Raymond Boisvert points out, the implicit hierarchies in Plato's philosophy restrict the good life to a small sector of society, granting "his successors a triple exclusion: banquets without food, men without women, and love without the body."[7] Deane Curtin also explores the ways that early philosophical considerations of food structured "what counts as a person in our culture," judgments based on conceptual dualisms that devalue the truth of everyday, physical experiences—mind over body, reason over emotion, self over other—and "buttresses the idea that one of the pair is the kind of thing that is autonomous and, therefore, fully real."[8]

Plato's student Aristotle gives us a slightly more complex picture of the role of the senses and their relation to the good life. Even though the pleasures of the senses can operate as a kind of virtue in the good life for Aristotle, it is still a matter of degree, as the lower senses in particular lend themselves to overindulgence.[9] One might be tempted to understand Aristotle's depiction of the good life and his principle of the "Golden Mean" as a version of the maxim "all things in moderation." And even though the midpoint certainly serves as a guide in the daily development of the moral virtues, the ultimate mark of Aristotle's good life retains Plato's bias against bodily pleasures and finds its fulfillment in the intellectual virtues and the pursuit of a highest good that is not simply valued in relation to some end, but is an end in itself.[10] Here the pleasures of the senses, particularly the lower ones, lack the connection to the arts, moral reflection, and the development of character.[11] One reason for this might be that cookery, as the preparation of a biological necessity for humans, is too end driven—as the art of cooking can only be enjoyed in its consumption, it cannot persist and lead to its reappreciation in the way the visual arts or a written work allow us to return again and again. Cookery, for Aristotle, seems to be nothing more than a technical skill rather than a form of practical wisdom.[12]

HIERARCHICAL AFTERTASTE

One of the more opportune chances for a vindication of gastronomical thinking arose in the seventeenth and eighteenth centuries in Europe when aesthetic philosophies of taste began to emerge. It is worth noting that aesthetics—the branch of philosophy that concerns questions of beauty and sensory experience—inherits philosophy's disdain for the body and the senses as sources of distraction away from rational contemplation. Not surprisingly, the philosophical discussion of beauty is thus shaped by the attempt to discover rational and universal measures for experiences that are unavoidably sensorial and subjective. So even though aesthetic taste becomes an issue for philosophy, "good taste" in matters of art and beauty dispenses with the physiological sense of taste and reasserts the themes of rationality, universality, and purity.

Korsmeyer details how the connection between good aesthetic judgment ("Taste") and gustatory experience ("taste") is lost due to the incompatibility of the latter with the prevalent aesthetic theories of the time.[13] For example, David Hume's "Of the Standard of Taste" (1757) resists the urge to let the subjectivity of individual taste preferences dominate his aesthetic theory (despite his considerable talents as a chef).[14] Hume argues that even though the standard of beauty or taste is found in the human subject rather than in the object of contemplation, he nonetheless insists on a common human element that allows humans to experience pleasure in a universal way. And while Hume does conflate the beautiful with the pleasurable, slightly opening the door for the appreciation of the culinary arts, he resolutely affirms that not everyone possesses the refinement of taste worthy of such judgments; thus, Hume's acknowledgment to aesthetic subjectivism is in the end consumed by traditional hierarchies of power.

In an attempt to overcome these subjective tendencies in aesthetic judgments, theories about the "disinterested" nature of aesthetic judgment emerge. To be disinterested in one's judgment means that one is free to contemplate the object at hand from an aesthetic standpoint without any ulterior motivations. Immanuel Kant, whose dinner parties were a hot ticket in Königsberg, nonetheless disqualifies the pleasures of food from aesthetic judgment and reinscribes taste within the ambits of rationality and universality.[15] In contrast to Hume, Kant completely dismisses the philosophical value of the pleasurable: a "pure aesthetic judgment" pertains only to beauty, is "disinterested," and thus has universal validity. A culinary creation, however pleasurable to the palate, in Kant's view fails to meet all three criteria.

Within the modern aesthetic tradition, food has little hope of being anything more than a pleasurable experience. But even by the standards of more contemporary aesthetic theories, the radical subjectivity of taste poses problems for the idea of "food as art." First and foremost, the sense of taste is rarely considered a "cognitive" sense, thus a culinary work of art can only appeal to the basest forms of pleasure, hardly warranting reflection in the way a great painting or a stirring dramatic performance might. The radical subjectivity of individual taste, crystallized in the motto *De gustibus non est disputandum* (There is no debating matters of taste), seems to cut short any discussion over meaning or authorial intent. Second, food art is unique because it is the only art form that doubles as a necessary means of biological survival: it fails to be an "end in itself"—or, more familiarly, "art for art's sake." Moreover, this act of consumption is not only fleeting (like a musical performance or improvised dance) but also, as the philosopher Hegel points out, literally destroyed in the process of its appreciation.[16] Some even argue that the use of food for art, or its enjoyment *as* an art, is a decadent and wasteful practice that reinforces social hierarchies of privilege and exclusion in the face of those who struggle to find sustenance.

Even contemporary philosophers who defend the importance of food in philosophical reflection grant only relative merit to the culinary arts. Elizabeth Telfer argues that food could certainly qualify as a decorative or minor art, but its restricted expressive range (pleasure/displeasure) and nonrepresentational nature (foods do not have intrinsic meaning) deny it access to one of the most basic functions of artistic expression: the creation of meaning through the representation of some other thing.[17] Korsmeyer claims that taste can be vindicated if it is shown to be already connected to cognitive and intellectual phenomena that have previously been reserved for sight and hearing—a concession to the aforementioned hierarchy of the senses. Particularly disturbing is Korsmeyer's support for representational food—namely, food that looks like other things, whether it be the crossed arms of a pretzel, a marshmallow Peep, a Kinder Egg, or even flavored gelatin molded into the bust of Elvis. This level of "artistic" engagement always risks falling back into the belief that food can be, at best, a minor or decorative art, and at worst, kitsch.[18]

The first philosopher to make "the pleasures of the stomach" central to his philosophy is Epicurus. Michael Symons argues that Epicurus's contribution to the philosophy of food is not so much the unbridled pursuit of pleasure so often associated with the term "epicurean," but a mindful, materialist philosophy well suited to a reconsideration of the value hierarchies that have dominated Western thinking.[19] Epicurus is the first to consider food positively in relation to the pursuit of the good life, and in a manner that was not completely at odds with the philosophical trends of his day. A radical reconsideration of the relation between the good life and the philosophical consideration of food, however, is still over two thousand years away. Brillat-Savarin's *Physiology of Taste* (1825) is subtitled "Meditations on Transcendental Gastronomy." While not generally considered a work of philosophy, his "transcendental" move suggests that "gastronomy" can be one of the necessary conditions that shape human reality. We should also not forget Jean-Paul Sartre, whose existential reflection on "the hole" saw in the act of eating a fundamental tendency in humans to fill voids and create meaning.[20]

But if one desires a crucial moment in the history of philosophy when food is redeemed, everything changes with Nietzsche: the question of the good life, the status of food and diet, everything. What is not certain, however, is whether Nietzsche's thoughtful consideration of human alimentation is a result of his radical reversal of the history of Western metaphysics and morals, or if his dietary interests are in fact the source of his "revaluating of all values." Nietzsche writes more consistently and deeply about food than any other philosopher in the Western canon; in this brief sketch, however, it will suffice to outline the manner in which he problematizes the persistent themes of rationality, universality, and purity.[21]

Nietzsche's struggles with illness throughout his life shape his writing and philosophy, and as was common during his time, he identifies the stomach and diet as the source of his physical (and metaphysical) issues. For Nietzsche, a healthy and vital digestive power involves a discerning openness to new ideas and an ability to incorporate those ideas in a process of becoming; in contrast to Plato's warnings about the appetitive soul, Nietzsche suggests that "'spirit' resembles a stomach more than anything."[22] Nietzsche problematizes the traditional hierarchies that place the soul above the body, reason above emotion, purity above plurality: "It is crucial for the fate of individuals as well as people that culture begin in the *right* place—*not* in the 'soul.' . . . The right place is the body, gestures, diet, physiology, *everything else* follows from this."[23] The gastrointestinal health of the body is directly related the spiritual health of an individual or society. Nietzsche thus asks: "Do we know the moral effects of foods? Is there a philosophy of nutrition"[24] to guide the growth of individual and shared life?

Nietzsche engages questions of food in order to explain his radical overturning of the history of morality. He does more than include food in his discussion of "the good life"; he uses the figures of digestion and nutrition to completely metabolize and transform the good life into the expression of our subjective tastes, an unapologetic confidence in one's own interests and creativity. In this discernment Nietzsche advocates neither the life of the hedonist nor the life of the ascetic. One has to understand how much and what sorts of food to ingest and, likewise, which virtues and ideas to accept and which ones to reject. To give voice to our tastes is to defend life itself, to defend the right that any one of us has to a life—based not on an abstract notion of universality but on the unique qualities of our particular life experiences.

For Emmanuel Levinas, the good life is similarly transformed; but rather than arrive at Nietzsche's radical subjectivity of taste, Levinas understands that food is the essential gateway to the ethical consideration of other human beings. In opposition to much of the tradition that precedes him, Levinas famously argues that "first philosophy"—the sort of philosophy that Plato and Aristotle attributed to pure essences, universal truths, and rational order—is not metaphysics but ethics. And rather than present the truths of first philosophy as ideas that could be cognized by a rational subject, Levinas construes ethics as an infinite responsibility to the other, one that resists rationalization and fulfillment. The other's demand upon us is visceral and direct: "To give, to-be-for-another, despite oneself, but in interrupting the for-oneself, is to take the bread out of one's own mouth, to nourish the hunger of another with one's own fasting."[25] David Goldstein's excellent treatment of food in Levinas's philosophy[26] not only places the philosopher at the center of any gastronomically sensitive account of the good life but also underscores the indifference (if not contempt) toward such themes by academic readers of Levinas. Food, and the fact that human subjects eat always in relation to other human eaters, is essential to the way that Levinas decenters ethics away from the individual will (central to Western ethics from Plato forward) and repositions it on the other's call to be responsible. Goldstein concludes that for Levinas it is "only by understanding what it means to eat can we understand the relation of the subject to otherness, and only a subject that understands what it means to eat can truly be *for*, not just aware of, the other. To be for-the-other is to know food's role in the constitution of the self. . . . Eating is *the* ethical event."[27]

THE THIRD MILLENNIUM'S THE CHARM

Despite the insights of Epicurus, Nietzsche, and Levinas, it would be fair to say that food only began to emerge as a compelling (but certainly not yet central) concern for philosophy in the wake of the various transformations that reshaped its theoretical landscape in the mid- to late twentieth century. As the hierarchies, dualisms, and essentialisms that characterized much of the history of Western philosophy buckled under the weight of historical change and sustained critique, so too did the prejudices that were formerly lined up against any philosophical interest in gastronomy. One could say that philosophy, in its many forms and schools, was slowly discovering new ways to temper, reinterpret, and in some cases abandon the singular pursuit of the rational, universal, and pure concepts so important to the tradition.

The critical moment in the philosophy of food appears to be the 1990s, a full decade ahead of the more popular "food revolution" that has, across all realms of American culture, transformed (and continues to transform) the way we think about food. It would be fair to say that the case for a philosophy of food began with the publication of *Cooking, Eating, Thinking: Transformative Philosophies of Food* (1992), by Deane Curtin and Lisa Heldke. The anthology collected older texts related to food but also included new essays designed to shape a new discourse around that tradition—one aimed at the transformation of the hierarchies that have marginalized and discounted the characteristics often associated with food: femininity, embodiedness, physical labor, desire, and animality. This transformation was achieved by directly engaging the tradition—reinterpreting some of its canonical authors, creating encounters with cultures and texts outside the Western tradition, and, perhaps most importantly, arguing that the pursuit of "the good life" in a philosophically responsible manner demands action: "We write from the suspicion of the theory/practice dichotomy, . . . in our commitment throughout the book to suggest concrete ways in which theoretical concepts do and should inform our actions in the world."[28]

In 1996, Elizabeth Telfer published *Food for Thought*, arguably the first monograph dedicated to the idea of a philosophy of food. Her premise is simple: eating and drinking are necessary for our physical survival; however, they also have value beyond mere feeding and watering.[29] Telfer is not so bold as to argue that her philosophy of food is ethically necessary, but she certainly leads the reader in that direction (particularly in her rejection of Plato's account of human nature). Her primary concern is that we do not overestimate the pleasure derived from food. This is indicated not only in her modest appraisal of food as a source of pleasure and an art form, but more importantly in the fact that two-thirds of the book are dedicated to the moral values associated with food; in particular, the book begins with a philosophical call to feed the hungry.

The aforementioned *Making Sense of Taste* (1999) is Carolyn Korsmeyer's powerful critique of sensory hierarchies within the philosophical tradition, one that not only vindicates the meaningfulness of our gustatory experiences but also places a more robust theoretical understanding of taste into contact with the visual arts, literature, and religious symbolism. In convincing fashion, Korsmeyer is able to show how taste is not a distraction from the life of the mind but, rather, a cognitive experience through and through that engages the most important philosophical questions.[30]

In the new millennium, many other books have followed. Notable are monographs by Michiel Korthals, Christian Koff, Rick Dolphijn, Nicola Perullo, and Francesca Rigotti, along with several anthologies dedicated to food and philosophy.[31] Raymond Boisvert's long-awaited reflection *I Eat, Therefore I Think* (2014) gives perhaps the most detailed account of the untapped gastronomical resources within the philosophical tradition to date, offering up "important reorientations in how we understand ourselves, our fellow human beings, and the natural world" in order to restore an "undersanding of philosophy as a 'way of life.'"[32] His collaboration with Lisa Heldke, *Philosophers at Table* (2016), further pursues this philosophical idea in a compellingly accessible format. But despite these offerings and the saturation of food-related media, the question remains: Is this "way of life" a necessary component of the good life? To conclude, I will sketch out how some of prevelant arguments suggest just such a responsibility.

If there is a manifesto for the philosophy of food, the best candidate is Lisa Heldke's essay "The Unexamined Meal Is Not Worth Eating."[33] A beautiful articulation of the reasons that philosophy could benefit from a serious engagement with food, the essay maps out the various avenues opened to and by philosophy when it takes everyday life, and our everyday life as eaters, as a starting point for philosophical reflection. Echoing my thesis in this essay, she states: "If philosophy takes seriously the claim that its questions are of fundamental importance for human life, then it has an obligation to explore those questions in ways that have at least the potential to engage people where they live—and eat. Food—a subject with immediate appeal, a subject with which everyone has some degree of familiarity, and a subject that carries its meaning on multiple levels—matters to people."[34] The good life is one in which we find better ways to coexist, or, as Boisvert often notes, ways to "con-vive"—to live *with*, literally, but also to share a table with others in the sense of a *convivium*.[35] Gastronomy represents one of the primary modes in which we live with others and relate to others; moreover, it is a mode of interrelation that is simultaneously linked to the biological need to feed our bodies. Food is not simply an object at our disposal; it is best understood, as Ileana Szymanski argues, as a set of active relations that bind us to each other and to the physical world we inhabit.[36] My suggestion is that these relations—and, more importantly, the good relations that, as Socrates notes in *Crito*, constitute our debt to the society that supports us—are not philosophy's goal but its necessary condition.

Only within the acknowledgment of this context—a horizon of human existence that encompasses body and mind unified in our existence as eaters—can philosophy understand the limits of its traditional themes and their relevance to "the good life." A philosophy of food thus proposes a new diet that does not abandon its traditional ingredients but strives to prepare them in new and inspiring ways. For example:

RATIONALITY. Philosophy has certainly not dispensed with reason, despite the challenges to its supremacy from within and without the discipline. Yet the philosophical consideration of food certainly helps to undermine its absolute supremacy as the means of accessing knowledge and truth. The works of Korsmeyer, Curtin, and Heldke help to dismantle the hierarchies—most notably that of mind over body—that have excised from human experience the rich context in which our ability to reason has meaning. More than just omnivores, we are omnivores whose choices need not be guided by instinct or situation alone but also by culture, passion, and creativity; the ability to use reason augments our opportunities for survival, of course, but also our pursuit of pleasure and happiness. This context includes not just our minds but our bodies that require and benefit from its nourishment, bodies that make the experience of food—as nutrient, and as bearer of meaning—different for everyone.[37] The rationalities of food express themselves through bodies, practices, objects, and systems that resist the simple test of the principle of noncontradiction. The good life is not simply ruled by reason but is open to a host of incommensurable logics while simultaneously granting supremacy to none of them.

UNIVERSALITY. Of course, philosophy traditionally employed reason in order to discover truths that would be universal and valid in every case. The encounter with food allows the philosopher to consider the variability of taste(s), on the side of the subject, and the food to be consumed, on the side of the object, as complex expressions of human existence rather than as threats to its universal definition. The expanding body of philosophical work on gustatory taste is the primary site for this revision of the tradition. And rather than simply fall into the subjectivity of individual tastes, the works of Korsmeyer and Perullo[38] detail the profound role that taste plays in human life as a fully cognitive experience and a source of existential meaning and critical insight. The good life is not simply a universal norm or ideal, but in fact a dramatic call to recognize and understand difference.

PURITY. Food is sometimes messy, and this has not sat well with the tradition of philosophy. But it does teach how illusory and overly abstract the traditional truths of philosophy have been. Nowhere is this clearer than in the debates over "authentic" foods,[39] and how the idea of an authentic, original, essential, and pure food or tradition is, like all claims to authenticity, problematic and fraught with caveats. This work is perhaps the most emblematic of the critical fruits that a philosophy of food can bear, as the careful exploration of claims to authenticity in food help to unravel how those same claims to purity shape (often in a negative and discriminatory way) social, political, and material relations. The good life is not a pure life, but one that understands the complexity and fluidity of relations that require continual maintenance, care, and conversation.

That emerging philosophies of food undertake a thorough and convincing critique of these traditional philosophical themes is vital not only to the development of the discipline of philosophy, but also to the human quest to live the good life. The good life certainly must involve food as a source of sustenance and nutrition; and the good life (rather than mere survival) likely also includes a level of pleasure derived from the consumption of food and from the social and cognitive relations that come along with it. But is a philosophy of food really *necessary* for the good life? Is a life that meets the goals of nutrition and pleasure still less than a flourishing, human life because it does not include thoughtful reflection upon the many different ways that humans interact with food? And is it then the responsibility of those who pursue such a life to ensure that those opportunities can also be enjoyed by others—that the pursuit of thought and reflection does not remain the privilege of the few?

One way to sort through these questions might be to reflect on the preparation of food, and in particular, the varying degrees of involvement that humans have had with it. To use a distinction in Aristotle's epistemology, the technical knowledge of cookery provides a convenient and apt counterpoint to the more theoretical wisdom that guides philosophy. In this collection, Albala's "A Plea for Culinary Luddism" and Laudan's "A Plea for Culinary Modernism" offer us a way to frame the issue. On the one hand, Albala points out that an ethos of scratch cooking might be the "key to our happiness," as the preparation, sharing, and enjoyment of food in this way is an "indispensable attribute of the good life" and a powerful antidote to the modern, industrial food system. His aim, as evidenced by his very personal relationship with cooking and food, is to inject a bit of lost wisdom into a foodscape dominated—as the philosopher Martin Heidegger might frame it—by industrial machination and the imperatives of technical efficiency, speed, and productivity. Albala's culinary romanticism is countered, however, by Laudan's pragmatic understanding of a more realistic gastronomical history—one that replaces the nostalgia for a bygone time with an appreciation for the safety, consistency, and diversity of a modern, global food system. While she admits that "we need to know how to prepare good food, and we need a culinary ethos," Laudan nonetheless urges us to practice an ethos that "comes to terms with contemporary, industrialized food" and the degree to which such a system, while not perfect, has increased culinary choice and freedom for a larger portion of human society. In both essays, we capture a glimpse of our problem's core: how might a reflection on our manifold relations with food—the practice of a culinary wisdom in the widest sense—lead us toward the good life, a process of self-actualization and flourishing? And how might such a life be made accessible to a greater number of individuals? Is merely engaging in the debate, weighing our options, and considering what it means to be an omnivore that produces its own food—individually or collectively—enough to help us become, as Nietzsche would say, who we truly are? What if the key to this debate is something we all possess, a capacity that is unique and universal at one and the same time?

To attempt an answer to this question, I will conclude with an idea that I use to begin my Food and Philosophy course for first-year students: the paradoxical certainty of individual taste. For them, philosophy is a big unknown. But what each of us likes, what tastes good to *me*, is usually as clear as day.

Our tastes are integral to our identities. Yet despite the physiological givens of our sensory organ of taste, this *experience* of tasting is not only cognitive but also, and perhaps more profoundly, a form of cognition shaped by the broader society in which we find ourselves. The source of our identity and uniqueness might turn out to be something that, as Levinas notes, arises in the others around us. And this suggests that taste could have deep ethical foundations.

Both Ludington ("The Standard of Taste Debate") and Finn ("Can 'Taste' Be Separated from Social Class?") recognize that our standards of taste, along with what is considered "good taste" (as compared to what "tastes good" in the most subjective sense), are socially constructed and performed in a manner that reflects hierarchies of class and status. Thus, the way that we as individuals generally experience taste should not be considered merely "subjective" but rather "objective" because it arises in conversation with the physical, conceptual, and social objects around us, and in turn, is itself performed as a social object.[40] Both authors are careful not to reduce this issue completely to social class; thus, a truly subjective act of taste, what "tastes good" to an individual, while shaped by social norms, cannot be eradicated by the force of social norms. We might surmise that there remains an aspect of subjective taste that does not enter the social sphere, or enters it only in the form of resistance and thus maintains a *critical* role (how else could tastes change?). What then empowers subjective taste, and even makes it *philosophical*—or, in other terms, a force of liberation?

Let me first point out the impasse to this realization, which takes two forms. The first form is the idea of a universal taste championed by eighteenth-century thinkers like Hume and Kant. As the chapters by Ludington and Finn suggest, the idea of a universal standard of taste is not an ideal that aspires to overcome the dangers of relativism; rather, it promotes the normalization of intellectual, political, and social hierarchies in the name of a "universal" humanity. Philosophically, it is the normalization of an idea of knowledge that, as I have already outlined above, is rational, universal, and pure. But the experience of taste—individually or socially—is nothing like that, and in this day and age, any whiff of universalism in the discussion of taste is quickly dispelled.

The second form is far more persistent, even though (I would argue) it is actually just a version of the first. Ludington expresses this form as "the need to remember that taste judgments aren't moral judgments." But what is this claim if not a (re)imposition of the intellectual hierarchies championed Kant, Hume, and in fact, most of the figures in Western intellectual history? Why are aesthetic judgments—shaped by social norms and practices, with real effects on the well-being of humans, animal and plant life, and the environment that sustains them all—distinct from moral ones? I would argue that aesthetic judgments are always already moral judgments precisely because taste always involves others and cannot be separated from its social impacts. The radical subjectivity of taste operates as a condition of possibility for that otherness—an otherness we all possess and that binds us to every other taster. If there is a universality operative in taste, it is not that pure science of a bygone philosophy; rather, it is one that is shared through its conversation with social norms and ideas and through its effects on those others who shape and are shaped by our aesthetic choices. In this way, taste is universal through acts of commensality.[41] Yet this universality is most fully expressed through radically subjective acts of gustatory resistance and the ultimate incommensurability of individual tastes. An open table is one that welcomes the stranger.

Such a table is not without risks: merely allowing the other a place at the table does not ensure productive dialogue or a functioning society. The noble risk of philosophy, recast in this gastronomical form, is wagered neither on a rational soul nor on universal ideas, but on the willingness of our rational self to be open to critique, to expose its own limits, and to allow its most basic physical and social needs a seat at the table when the question of the good life is up for discussion. What therefore matters philosophically is the fact that we all have individual tastes, the fact that they are to a great degree unique and incommensurable, and, most importantly, that this fact grounds the right each of us has to our own taste: a right that is tied to every other taster. This is not a rational ideal, but one that is enacted every time we employ our sense of taste, every time we find pleasure in tasting. If we accept that no individual can tell any other individual that a taste should please or displease them, we find ourselves in possession of a powerful foundation for an inclusive society. What makes us equal is not an abstract right or notion of reason, but the lived experience, which we all share in common, of knowing what we like. Could our responsibilities to others be based on the simple fact that every person has the right to their own preference; every person has a voice; every person deserves a seat at the table?

Building on many of the philosophies of food mentioned above, this could certainly serve as a *critical* principle that gives voice to those who have, through traditional hierarchies of value and status, been denied equality of opportunity or excluded altogether. But could it also be a constructive principle on which the good life could be imagined—a good life not only for individual tasters, but for a community of individual tastes? Could we ever find common ground? We are left to consider what the good life might be if it is measured according to criteria other than rationality, universality, and purity—criteria that have more often than not been used as tools of exclusion. Not that their opposites necessarily ensure a better society, as irrationality, particularity, and contamination are an equally frightening prospect to many philosophers, let alone to the majority of individuals who understand the value of good sense, equality, and the power of ideals. To pursue these questions—there at the point of convergence between body and mind, individual taste and social responsibility, human nourishing and human flourishing—is the critical work of a philosophy of food and, subsequently, the pursuit of the good life. If, however, the reader of this essay chooses to reject my initial presuppositions and politely passes on what Socrates describes as "the examined life," all is not lost. There is still much to be gained, outside of philosophy, when we sit down at the table together to share a meal.

NOTES

1. The meaning and seriousness of Feuerbach's phrase—which in German includes a play on words, "Der Mensch ist [is] was er isst [eats]"—has been the subject of some disagreement. Cf. Melvin Cherno, "'Man Is What He Eats': A Rectification," *Journal of the History of Ideas* 24, no. 3 (1963): 397–406.

2. Jean Anthelme Brillat-Savarin, *The Physiology of Taste* (Harmondsworth, U.K.: Penguin Books, 1994), 84; translation modified.

3. Lisa Heldke details the various ways that philosophy can approach food issues in "The Unexamined Meal Is Not Worth Eating: or, Why and How Philosophers (Might/Could/Do) Study Food," *Food, Culture and Society* 9, no. 2 (2006): 201–19. Along with Peter Singer, Paul Thompson has perhaps been the most prolific philosopher in the realm of applied ethics, particularly related to agricultural issues. See especially *The Spirit of the Soil: Agriculture and Environmental Ethics* (London: Routledge, 1994) and *The Agrarian Vision: Sustainability and Environmental Ethics* (Lexington: University Press of Kentucky, 2010).

4. Food has certainly been present, and at times central, to the subjects of the dialogues, but its role is not often positive. Cf. Carolyn Korsmeyer, *Making Sense of Taste: Food and Philosophy* (Ithaca, N.Y.: Cornell University Press, 1999); Lisa Heldke, "The Unexamined Meal Is Not Worth Eating," 202; Raymond Boisvert, *I Eat, Therefore I Think: Food and Philosophy* (Madison, N.J.: Fairleigh Dickinson University Press, 2014); and Raymond Boisvert and Lisa Heldke, *Philosophers at Table: On Food and Being Human* (London: Reaktion Books), 2016.

5. Plato, *Plato in Twelve Volumes*, vol. 1, trans. Harold North Fowler; introduction by W. R. M. Lamb (Cambridge, Mass.: Harvard University Press; London: William Heinemann, 1966), 64d.

6. Korsmeyer, *Making Sense of Taste*, 14.

7. Boisvert, *I Eat, Therefore I Think*, 11.

8. Deane Curtin, "Food/Body/Person," in Deane W. Curtin and Lisa M. Heldke, *Cooking, Eating, Thinking: Transformative Philosophies of Food* (Bloomington: Indiana University Press, 1992), 4–5.

9. Korsmeyer, *Making Sense of Taste*, 21–22.

10. Aristotle, *Nicomachean Ethics*, book 1, trans. and with introduction and notes by C. D. C. Reeve (Indianapolis: Hackett, 2014).

11. Cf. Korsmeyer, *Making Sense of Taste*, 24; Boisvert, *I Eat, Therefore I Think*, 76.

12. This distinction, from Aristotle's *Nichomachean Ethics*, book 6, is one that is revisited by numerous scholars.

13. Carolyn Korsmeyer, *Making Sense of Taste*, esp. chap. 2, "Philosophies of Taste: Aesthetic and Non-Aesthetic Senses," 38–67. Similar themes and figures are addressed in Kevin W. Sweeney, "Can a Soup Be Beautiful? The Rise of Gastronomy and the Aesthetics of Food," in *Food and Philosophy: Eat, Think and Be Merry*, ed. Fritz Allhoff and Dave Monroe (Oxford: Blackwell, 2007), 117–32.

14. Boisvert, *I Eat, Therefore I Think*, 14.

15. Immanuel Kant, *Critique of Judgment* (Indianapolis: Hackett, 1987).

16. Hegel states that "we can smell only what is in the process of wasting away, and we can taste only by destroying." G. W. F. Hegel, *Aesthetics: Lectures on Fine Art*, trans. T. M. Knox, 2 vols. (Oxford: Clarendon, 1975), 1:137–38. For a defense of food as art in light of its unavoidable consumption, see Dave Monroe, "Can Food Be Art? The Problem of Consumption," in Allhoff and Monroe, *Food and Philosophy*, 133–44.

17. Elizabeth Telfer, *Food for Thought: Philosophy and Food* (London: Routledge, 1996), 41–60.

18. Korsmeyer, *Making Sense of Taste*, 118–28.

19. Michael Symons, "Epicurus: The Foodie's Philosopher," in Allhoff and Monroe, *Food and Philosophy*, 13–30.

20. Jean-Paul Sartre, *Existentialism and Human Emotions* (New York: Philosophical Library, 1967), 84–90.

21. For more comprehensive account of food in Nietzsche's philosophy, I refer the reader to Robert Valgenti, "Nietzsche and Food," in *Encyclopedia of Food and Agricultural Ethics*, ed. Paul Thompson and David Kaplan (Dordrecht, Netherlands: Springer, 2014).

22. F. W. Nietzsche, *Beyond Good and Evil: Prelude to a Philosophy of the Future* (Cambridge: Cambridge University Press, 2002), 122.

23. F. W. Nietzsche, *The Anti-Christ, Ecce Homo, Twilight of the Idols, and Other Writings* (Cambridge: Cambridge University Press, 2005), 221.

24. F. W. Nietzsche, *The Gay Science: With a Prelude in German Rhymes and an Appendix of Songs* (Cambridge: Cambridge University Press, 2001), 34.

25. Emmanuel Levinas, *Otherwise Than Being: or, Beyond Essence*, trans. Alphonso Lingis (Pittsburgh: Duquesne University Press, 1998), 56.

26. David Goldstein, "Emmanuel Levinas and the Ontology of Eating," *Gastronomica* 10, no. 3 (2010): 34–44.

27. Goldstein, "Emmanuel Levinas," 42.

28. Curtin and Heldke, *Cooking, Eating, Thinking*, xiv.

29. Telfer, *Food for Thought*, 1.

30. See also Carolyn Korsmeyer, "Delightful, Delicious, Disgusting," *Journal of Aesthetics and Art Criticism* 60, no. 3 (2002): 217–25.

31. In chronological order: Francesca Rigotti, *La filosofia in cucina* (Bologna: Il Mulino, 1999); Michiel Korthals, *Before Dinner: Philosophy and Ethics of Food* (Dordrecht, Netherlands: Springer, 2004); Rick Dolphijn, *Foodscapes: Towards a Deleuzian Ethics of Consumption* (Delft: Eburon, 2004); Christian Coff, *The Taste for Ethics: An Ethic of Food Consumption* (Dordrecht, Netherlands: Springer, 2006); Nicola Perullo, *Il gusto come esperienza* (Turin: Slow Food Editore, 2012). Recent anthologies include Gregory E. Pence, *The Ethics of Food: A Reader for the Twenty-First Century* (Lanham, Md.: Rowman & Littlefield, 2002); Allhoff and Monroe, *Food and Philosophy*; David M. Kaplan, ed., *The Philosophy of Food* (Berkeley: University of California Press, 2012); and special issues of two academic journals: *Collapse: Philosophical Research and Development* 7 (2011) and *PhaenEx* 8, no. 2 (Fall/Winter 2013).

32. Boisvert, *I Eat, Therefore I Think*, 2.

33. Heldke, "The Unexamined Meal Is Not Worth Eating," 201–19.

34. Heldke, "The Unexamined Meal Is Not Worth Eating," 203.

35. Boisvert, *I Eat, Therefore I Think*, 18.

36. Ileana F. Szymanski, "The Metaphysics and Ethics of Food as Activity," *Radical Philosophy Review* 17, no. 2 (2014): 351–70.

37. On the recognition and critical application of these varying rationalities, I especially recommend the following: Dolphijn, *Foodscapes*; Jane Bennett, "Edible Material," in *Vibrant Matter: A Political Ecology of Things* (Durham, N.C.: Duke University Press, 2010), 39–51; Allison Hayes-Conroy and Jessica Hayes-Conroy, "Taking Back Taste: Feminism, Food and Visceral Politics," *Gender, Place and Culture: A Journal of Feminist Geography* 15, no. 5 (2008): 461–73.

38. Nicola Perullo, *Taste as Experience: The Philosophy and Aesthetics of Food* (New York: Columbia University Press, 2016).

39. The topic of authenticity has proved rich and complex for a range of thinkers. See in particular Andrea Borghini, "What Is a True *Ribollita*? Memory and the Quest for Authentic Food," in *Segredo e memória: Ensaios sobre a Era da Informação* [Secret and memory in the Information Age], ed. T. Piazza (Porto, Portugal: Porto, 2010); Andrea Borghini, "Authenticity in Food," in *Encyclopedia of Food and Agricultural Ethics*, ed. Paul Thompson and David Kaplan (Dordrecht, Netherlands: Springer, 2014), 180–85; Lisa Heldke, "But Is It Authentic? Culinary Travel and the Search for the 'Genuine Article,'" in *The Taste Culture Reader: Experiencing Food and Drink*, ed. Carolyn Korsmeyer (New York: Berg, 2005), 385–94; Fabio Parasecoli, "Tourism and Taste: Exploring Identities," in *Bite Me: Food in Popular Culture* (New York: Berg, 2008), 127–45; and Sharon Zukin, "Consuming Authenticity," *Cultural Studies* 22, no. 5 (September 2008): 724–48.

40. Here I do not intend "objective" in the sense of naive or materialistic realism. "Objective" in this sense means that the causes, and also the effects, of a certain experience or viewpoint cannot be reduced to the individual subject.

41. "Commensality" is of particular importance in the work of Boisvert and Heldke.

12

A Plea for Culinary Luddism

KEN ALBALA

The importance of cooking and eating together has become a heated issue among people of every political persuasion. For some, the past century has been a triumph in providing adequate food and nutrition to a rapidly growing population despite a number of lingering inequities. For others, the food industry and convenience foods have robbed us of the benefits of cooking from scratch, sharing food, and eating together. This essay takes the latter position: not only is cooking central to what it means to be human, but cooking from scratch using whole ingredients and gathering around the dinner table may itself be essential to our happiness.

It should be clear from the outset that this is a personal cri de coeur rather than a research article. It was conceived partly in response to Rachel Laudan's now-classic essay in which the term "food Luddite" was first coined. I harbor the illusion that Laudan's term was devised in reference to me. Whatever the case, it was also clearly intended to resonate with the works of popular writers like Michael Pollan and Mark Bittman, both of whom made valiant calls to reclaim the kitchen, although not in ways that appeal beyond the class of leisured elites for whom cooking has become a fashionable hobby. Moreover, my plea is not made primarily in the interest of health, the environment, animal welfare, or the host of other food issues now mustered in defense of home cooking. Nor is my intention to wag a finger at people supposedly too lazy to cook; I recognize the myriad social pressures that have steered us away from the kitchen. I argue that cooking is fundamentally a matter of pleasure, akin to that found in other pursuits like dancing, singing, playing sports, and telling stories, all of which I believe helped us evolve as social beings.

I am often asked why I spend so much time in the kitchen laboring over antiquated techniques; why I prefer cooking from scratch; why I like preserving food with bygone methods and fiddling with manual contraptions that no one in their right mind would use anymore when we have so many modern conveniences. This essay is an attempt to answer these questions from a purely personal angle, but it also suggests why cooking is so important, in fact central to our humanity. Simply put, I can answer Bob Valgenti's question in chapter 11: to prepare food, to share it, to get pleasure from the act, is an indispensable attribute of the good life. The modern food industry, wittingly or not, has done everything in its power to rob us of the pleasure of cooking and eating and sharing food. I do not propose turning back the clock, and I suffer no delusions about how difficult life was in the past, especially if you were poor or a woman. I understand Rachel Laudan's and Tracey Deutsch's arguments in their chapters, but I do think that we have gone too far trusting modern industry to make our lives better and that the kitchen is among the most valuable places in which we can devote our time and energy.

The appeal of cooking from scratch with fresh ingredients is partly a cyclical phenomenon; it has gone in and out of fashion in the past century. Every generation to some extent rejects the values of its parents. People who grew up in the city longed to escape to suburbia. Their children couldn't wait to get back to the city. The same is true of food. For the generation that grew up in hard times, especially the 1930s, convenience food and culinary short cuts were a godsend. Their children, once they grew up, looked on these industrial foods with contempt. So it was in my family. My mother grew up in Depression-era Brooklyn. To save two cents her mother would buy chickens with feathers still on and pluck them at home, and she often got ticks from them. She served her children lungs and liver and all manner of cheap "variety meats." I am told my mom and her siblings liked them until they got older and figured out what they were. For my mother's generation escaping hard times simply meant not having to slave in the kitchen, and in the 1950s they happily welcomed every single convenience food dreamt up by scientists. They loved cooking from cans, and for them Poppy Canon's *Can Opener Cookbook* (1952) was the best thing since sliced bread. Children of the Depression were also permanently influenced by the apprehension of hunger. My mother would hoard food and had a refrigerator and a separate freezer in the garage filled mostly with almost stale bread which she defrosted for company and declared "just like fresh!" She certainly never wanted for food as an adult, but somehow the idea of an empty cupboard terrified her. I would call it gastronomic horror vacui: having a lot of bad food is better than having none at all.

When I was young, in the 1960s and 1970s, health food and natural food became popular. *The Moosewood Cookbook* came out in 1974, and Mollie Katzen admitted that the whole book was a reaction to the food she grew up with. Mainstream Americans ate a lot of junk. I certainly did. I was plied with devil dogs and potato chips, and I still have a terrible weakness for chips. The cabinet was always crammed with snacks, and there were secret stashes of candy and cookies all over the house. The living room had a huge basket filled with stale bargain store junk food. I ate it happily as a kid. As long as I had my can of Campbell's chicken noodle soup in a thermos and a peanut butter sandwich on white bread in my Snoopy lunchbox, I was good to go. Because my father commuted to the city and came home late, I remember eating dinner on my own in front of the TV on a small foldable metal tray table. They were called "TV tables"; I don't think they even exist anymore. My dining companions were Gilligan, Herman Munster, and Samantha Stevens. I have only the vaguest recollection of what I was fed, but it was never anything interesting. For my mother cooking was a chore. The quicker it was over with, the better. TV dinners were a blessing.

Perhaps in reaction to this upbringing it is no wonder that I gained deep interest in cooking. Perhaps I inherited a cooking gene from my paternal grandmother. She died when I was thirteen, but my only memory of her is in the kitchen. At the slightest suggestion that someone wanted something to eat, she would be busy, cigarette dangling from her mouth, draped in a housecoat, with her wobbly fat arms flailing about, rolling out dough, chopping vegetables, or making candy. Some of the flavors of her cooking I remember with such profound precision. My grandmother was Sephardic, so she made things like Boyos, which were little oily rolls filled with spinach and feta; susam was hard sesame candy; macaroni reynado was a lush baked casserole with ground meat and a tomato-based sauce. It was clear that the greatest pleasure in life for my grandmother was feeding those she loved, and everybody else too. I remember sitting on a red stepstool watching her intently as she cooked, and now only regret that I never wrote anything down. Not that cooking for her was a choice, as with most women of her generation, but she embraced it wholeheartedly.

I never got to cook consistently until I went to college, but I do remember various odd projects in my youth. I used to make blueberry wine at summer camp in plastic pitchers. I had a firepit in the backyard for cooking: one day it might be a sour cherry pie with fruit from the trees, or a stew with wild onions that grew in the lawn. When I was unleashed in an apartment kitchen in college, I never looked back, never even thought twice if I could make a dish, but jumped in with gusto. I am certain that I learned everything I needed to know from TV. There was "The Galloping Gourmet" and Julia Child, a little later Martin Yan, Nathalie Dupree, Justin Wilson,

Jeff Smith. Above all Jacques Pepin's knife skills were enthralling. For the past thirty-something years, scarcely a day has gone by without me cooking something. It's an addiction, but food has become my profession and my passion. I like nothing better than feeding a big crowd, reading cookbooks, teaching about food, and writing food books. I realize this is not ordinary behavior; I am simply trying to convey my enthusiasm. Nor am I suggesting that good food must be complicated food—exactly the opposite.

So perhaps you will understand why I react with shock and amazement when someone tells me, "I just don't like cooking." "It's so much work!" I had a heated argument one day with a woman who insisted that some people like sports, some people like music, and some people like cooking. Why should anyone be forced to do something they don't like when there are frozen dinners, takeout, and decent restaurants that can do all the work for you? Many arguments can be mustered on my side. Cooking from scratch is better for you, it's almost always less expensive, and it usually tastes better. But ultimately what it really boils down to, at least for me, is that it's pleasurable: the experimentation, learning to use new ingredients, combining flavors in new ways, and especially tackling things that no one thinks can be done anymore, like baking fresh sourdough bread, curing meats, distilling alcohol. The perfect day for me is having spent hours facing the stove, experimenting, creating, and then eating. It was recently pointed out to me that I live in the kitchen and stand at the cutting board, even when not cooking. Quite simply, for me, cooking and even going through the labor of grinding, soaking, fermenting, curdling, smoking is not work, it's play. It's a way to unwind and do something both creative and useful.

The ultimate pleasure is of course not the cooking itself but sharing what you have cooked and watching people enjoy it. Really, what do people want in life other than the esteem of others, having people know that you spent time in the kitchen to make them happy? Of course everyone can't be cooking all the time, but if everyone gave it a shot on occasion, or shared kitchen work, it would never seem like a burden. The esteem could be shared among everyone who takes a turn cooking. Since we have been gradually moving toward gender equity in the household, this activity should be shared, like all others, though again not seen as a chore. It only becomes a chore, as Tracey Deutsch points out in chapter 10, when it is only left to one person in the household, usually a woman.

Let us consider why we do anything in life: collect possessions, excel in some activity, or gain some particular expert knowledge? As I've already argued, one reason we do these things is because we seek praise from others.

But there are other reasons as well. And the question remains whether preparing food must be one of these activities. And why does it need to be from scratch? And why for heaven's sake, should we use antiquated methods? Why would anyone in their right mind grind ingredients by hand when there are food processors? Why wait patiently for a stock when you can open a can? Why cure salami for three months when you can buy one at the store? Or make moderately passable wine when you know the experts are better at it?

Part of my answer to these latter questions is a consequence of my interest in the past. I am a food historian, so that using old techniques is a form of research. I want to know how things were done before the industrial era. What kind of muscles were involved in cooking over an open hearth? What kind of bacteria did people harness to make pickles that would last them through the winter? How did they manage a wood-fired bread oven? I will readily admit, these techniques can be difficult and dangerous. They are not the kind of things one would want to do on a daily basis unless it were a full-time job. Even then I am certain it would get very tedious if one's life depended on it, as of course it did in the past, especially for working classes. Again, I am not advocating turning back the clock. Modern conveniences have freed up time so we can pursue other activities. We can now spend more time at work and more time in front of the TV or computer.

Indeed, what have we lost by having forgotten antiquated techniques? A lot, actually. Few of us have any idea where our food comes from anymore. Few people have butchered an animal or separated curds and whey. Most people haven't experienced the thrill of opening a jar of homemade soy sauce after a year of letting it ferment on the shelf. But does that really matter? Why not leave such things to professionals and food scientists? Should we not trust them the same way we trust people who make bridges, design our cell phones, and spend our tax dollars?

Here is where I must make a plea for culinary Luddism and make use of a few familiar arguments. The food industry purports to be making our lives easier and more enjoyable, but is it to be trusted? Industrial agriculture has wrought havoc on the environment and decimated rural communities. It has paid poor wages to workers, with especially disastrous consequences for people in the global South. It has subjected animals to unspeakable horrors. And it has made whopping mistakes when it comes to human health. My parents' generation was made to revile lard and even butter in favor of shortening and margarine. Now we know that trans fats are just about the worst thing you can put in your body. Nutritional science remains fickle for many complicated reasons, as Charlotte Biltekoff points out in chapter 6, but the fact remains that much of the nutritional advice with which the public is assailed is designed simply to sell more industrial food.

It is clear that the food industry has profit in mind, not human happiness. Even more frighteningly, processed food is profoundly different than food cooked with whole raw ingredients, solely from the perspective of flavor. A huge array of artificial and so-called "natural" flavors and fragrances go into processed food to boost the flavor. Doritos, their flavor boosted to the nth degree, are an excellent example. These are designed to hit the palate hard and then immediately fade, so you are goaded into eating more. In other words, these foods are designed to make us obese. And our government subsidizes them by paying farmers to grow corn and soy—exactly the industrial products that go into mass-produced junk food. And no wonder children don't like fruits and vegetables when their taste buds are assaulted by things like Doritos and Flamin' Hot Cheetos. We are raising a generation whose palates no longer perceive subtlety and nuance in food, can no longer appreciate a ripe peach, not only because they're shipped across the country out of season but because people literally cannot taste them anymore.

The act of cooking enhances the pleasure of eating. Much of modern entertainment is passive. We watch people cook on TV, we watch people play sports, or watch rock stars perform. In the past the pleasure came from doing and making things yourself, or having people in your immediate circle do these things. You cooked dinner, went outside to toss a ball, or scratched away on the fiddle yourself. It may not have been professional, or commercialized the way the modern entertainment industry spoons it out to us, but it was real. Of course, wealthy people could go to restaurants or hire their own court composer, but most people had to provide entertainment and sustenance themselves. There can be no doubt that playing a sport, performing with others, and cooking (also with others) is infinitely more rewarding than switching on the TV, radio, or microwave. This is not romantic nostalgia for a preindustrial or precapitalist past—it is a program for the future, a way to become less dependent on corporations whose professed goal is making us happier by making our lives more convenient, but whose real interest is reaping profit by blinding us to more profound forms of happiness. Cooking is a creative activity we can all partake in, and it is a way for us to take control of our lives, which are increasingly alienated from the very activities that make us happy. In other words, while cooking might feel like drudgery to some, it is an activity that we forsake at our peril. Cooking good food, with and for family and friends, nourishes our bodies and our souls.

But if I am correct and cooking is both fun and fulfilling, then why do so many people leave the task to others? I suspect it is largely due to intimidation and time. Watching expert chefs compete on TV with competing with outlandish ingredients teaches very few practical cooking skills; in fact, I suspect it discourages many people from even trying. Moreover, basic skills have been lost, and this was not by chance. The food industry makes more money when they do the work for you. It's not merely the prepared food you find in the central aisles of every supermarket—the snacks, cans, and frozen dinners. You can also buy precut vegetables, cake mixes, premarinated meat, even a cut-up chicken, which is about twice the price per pound of a whole bird. Have we just become so lazy that it is worth the extra money (or rather time spent at work to earn that money), so we have more shortcuts?

I understand that the situation is more complicated than that. We seem to be convinced now that there is no time to cook. For many people, grabbing whatever is at hand seems to be more convenient. There's commuting time, an hour at the gym, shuffling kids to activities, cleaning the house, and dozens of other things we need to do, or feel we need to do. It only makes sense to outsource some of these tasks so we can do other things we like. In short, lack of time seems to be the most common excuse for not cooking on a daily basis. And really, who enjoys cleaning up and doing dishes? Takeout saves you that drudgery. Disposable containers and plastic cutlery just go in the trash.

But all of this convenience comes at a much higher cost than the time it takes to cook, eat, and clean up a decent meal. The garbage accumulated by our modern food system is deplorable, but even more serious is the waste. A UNEP and World Resources Institute report suggests that one-third of all food worldwide is wasted either in the field or in transport, retail, or consumption.[1] In the United States this comes to twenty pounds of food per person every month. Much of what is grown never makes it to the shelves because it is spotty, imperfect, or for some other reason deemed unfit for picky consumers. Apparently people prefer a beautiful Red Delicious apple than one with good flavor. We must learn to love less-than-perfect produce. We must remember how good the odd parts of animals taste, the guts that I was never served as a kid. They have been finding their way into high-end restaurants, to be sure, but until they end up on people's plates instead of in dog food, we are wasting so much money rearing animals only to eat a few choice, tender muscles. This is certainly a lesson we can learn from the past—waste nothing. I hold the food industry responsible for our modern squeamishness: they want to sell only the most expensive cuts. Now and then some animal part will become fashionable and they are happy to raise the price—think of soup bones now that bone broth is touted among the Paleo diet crowd. When this same product was called *stock*, butchers would often give away bones for free.

As you will have noticed, a large part of this essay revolves around the concept of flavor. Setting aside all the environmental, social, political, and ethical problems caused by our modern food supply, I think the one usually given shortest shrift is simply that most industrial food doesn't taste that good. Cooking yourself and even once in a while expending enormous amounts of labor is not only fun and fulfilling, but the taste of the food is also inherently superior. Many will argue, as Margot Finn does in chapter 4, that taste is subjective. And how can I possibly know food that is grown or made in labor-intensive ways is inherently healthier? Look on the positive side, says Peter Coclanis in chapter 2; our food supply is by and large safe. True. We grow enough food with modern agriculture even if it isn't equitably distributed and not everyone has access to it. True. And we are healthier, taller, and live longer than in the past. True again. Why then would anyone want to learn things that our ancestors happily abandoned? My answer: it is mostly a matter of superior taste. I need not have you taste a real naturally fermented cheese next to a rubbery industrial one. Nor a beautifully baked artisanal loaf of bread next to a faux-rustic supermarket loaf. Or even a gorgeous hoppy craft beer next to a can of Budweiser. There are very good reasons these artisanal products are gaining in popularity: they taste great. I would go even farther than Charles Ludington does in chapter 5 and assert unequivocally that some things are objectively better-tasting than others. Nor are many of these products, despite what their advertisements might claim, made using historic traditional methods. Instead, they are as much the product of modern science as a box of Froot Loops is. No serious commercial brewer would let his beer be subject to the vagaries of capricious yeast the way it was done in the past.

Then again, you will ask, why should we not leave all these difficult processes to the professional artisan rather than try it at home? Because with all these artisanal products, as fine as they might be, you are still a passive consumer. With the greater scale that comes with commercial success for artisan businesses, inevitably corners are cut, shelf life becomes a consideration, industrial protocols and government regulations must be followed. Not to mention that the craft aesthetic mostly caters to elites and usually costs more than industrial food: one more good reason to cook at home.

There is something to be said for the small and free space of your kitchen as far as quality is concerned. But more importantly, making these products yourself channels your creative energy in way that allows the product to become an extension of oneself, much as Marx argued in his critique of capitalist labor. The value derives not from a price tag but from the pleasure of having made it with your own hands. It is a simple matter of pride, the joy of sharing it with others, and having mastered a technique. Things do sometimes go wrong. There can be horrendous failures—unrisen bread, moldy pickles, sour wine. But when it turns out right, the flavor can be incomparable. The serendipitous success is worth more than anything that can be purchased. That is precisely what we have lost from the past—the pleasure gained from something truly homemade. And admittedly people in the past bought bread and wine and much else; self-sufficiency has never been the norm in human history, nor need it be today. It is simply that these are things we can indeed do at home, even if only on occasion, and are worth doing if only to avoid the disgraceful dependence on industrial pabulum.

Tied into this entire discussion is the supposed value of the family meal. It is a topic heavily laden with political rhetoric. The home-cooked meal is seen by many as the foundation of the traditional American family, whether such a thing really ever existed or not. We historians know that the idealized American conjugal unit and nuclear family was really only a construction of the nineteenth century, more what people hoped would happen than a reflection of reality. All the same, Janet Flammang's book *The Taste for Civilization* argues that the family meal is the practical arena for teaching lessons in civility, where children learn to discuss problems, reach compromises, and ultimately participate in democracy.[2] The table is where we learn to listen to others, to sympathize with them, and sometimes when to keep quiet. Because we are political creatures, dinner is the most important place we engage with our companions—the root of the word "companion" itself means people with whom we share bread. Moreover, ritual meals reinforce kinship relations, allow people to reconnect with relatives, and help assert ethnic, cultural, or religious identity in an increasingly impersonal and lonely world.

Eating together is difficult. The demands of modern life, hectic working schedules and commuting, and children's activities all force us to eat on the go, or eat out on a frequent basis. Cooking is relegated to special occasions, as a recreational activity for the wealthy educated classes rather than an everyday necessity for everyone. Every year fewer people actually know how to cook. Surveys on this topic vary widely since few people can agree on a common definition of cooking, but subjects who themselves admit they don't know how to cook usually range from 25% to 30%. Every study has shown that people consistently cook less every year, despite the current obsession with food.[3] This has had a direct impact on the frequency of family meals, however one might define a family and whatever the supposed benefits socially, culturally, or politically.

But what really goes on at the dinner table when we do sit down together? As much as we like to imagine civil conversation among family members, realistically the table is often the site of intense contention. It can be an emotionally trying experience. A meal can be a battlefield where everyone vents daily frustrations, picks on siblings, or allows conjugal animosities to seethe. Many people prefer simply to opt out. Large gatherings can be worse, when long-standing arguments resurface. As Alice Julier's book *Eating Together* shows, common meals are as much about drawing boundaries and excluding others as they are about bringing people together.[4] Even if the atmosphere remains cordial, it can be and often is interrupted by cell phones, the TV, people coming and going, and what seems to be a rush to just get the whole thing over with as soon as possible.

Complicating the situation, few people want to eat the same thing anymore, making planning a logistical nightmare. Diets, allergies, aversion to meat, gluten intolerance, lactose intolerance, faddish preferences, and plain old pickiness demand a staggering array of dishes, even for a small gathering. Every parent has experienced the disappointment of children rejecting the dinner you lavished your efforts on cooking. The days of one main dish and a few sides are gone. Sometimes it is just easier to let everyone fend for themselves, even if it means everyone eats something different, eats by themselves, or just grazes on junk.

Therefore, the question remains: Is commensality really worth the bother? Who needs to cook a family meal when we have so many other options? Should we just let the common meal vanish and place our gastronomic fate in the hands of corporations? The food industry benefits from our social anomie by offering ready-made frozen dinners that can be popped in the microwave plus a wide variety of fast food and casual dining chains, increasingly with drive-through service, that allow people to eat whatever they like, whenever they want it, and, most importantly, alone.

The common meal is clearly in danger of extinction, yet again I propose one good reason that might motivate us to endure the ordeal of cooking and eating together: a more profound form of pleasure that comes with practice and time. Let us assume for the moment that the ancient philosopher

Epicurus was right when he said that humans are essentially selfish, calculating creatures. We never do anything unpleasant unless we anticipate greater returns. In our heads we run a simple cost-benefit analysis, trying to avoid pain and maximize pleasure in every decision we make. This is not the caricature Epicurean who simply seeks rarefied pleasure as the index of happiness—exactly the opposite. Sometimes we consciously limit immediate pleasures in hope of deferred benefits, or forgo particular pleasures if they bring greater suffering in their wake. Of course, we don't always know what is in our own best self-interest, and here, I would argue, is exactly where we stand on the issue of commensality.

We simply do not recognize the benefits of cooking and sharing food that make it worth the effort. And here I return to my initial argument about the importance of cooking. The benefit of cooking on a daily basis is that we receive the love and appreciation of those for whom we've cooked. This is the most profound thing we humans want in life. It is not merely the company of others but the admiration of others. We seek social standing or to put it more bluntly, applause. Some people accumulate money, but isn't it ultimately to buy objects to fetishize, which we imagine will impress people? It is still all about the praise of others, something we need as human beings. On some level we all want to be rock stars. Most of us can't be. However, we can all be cooks.

When we care about cooking, we cook well. And when we cook well, we reap the praise of our friends and family members. At dinner parties we get compliments. From strangers we earn accolades. Even bloggers and Facebook addicts ultimately get a rush from comments and "likes" even though, ironically, no one online actually tastes the food they cook. Cooking and sharing food are inseparable. Mere eating together is something entirely different. Although we might enjoy that bucket of takeout chicken, it simply does not yield the personal pleasure one gets from cooking and feeding others, nor the pleasure of knowing that someone personally attended to your nourishment and well-being.

Our labor in the kitchen congeals in the form not of profit but of praise. That's the only reason we do anything, and it makes sweating in the kitchen for hours ultimately worth it. It motivates us to spend months perfecting a difficult lick on the guitar, training to break a record marathon time, building a beautiful cabinet. We might make money from these pursuits, but the real pleasure is in the actual doing and the reaping of esteem.

In that regard, cooking only for ourselves is not the point. Instead, we should cook for others and eat with others. There is nothing more important than this, for the simple reason that we need to eat every day, so we might as well reap the greatest benefit we can. It is the most immediate and palpable pleasure a human being can invest time in—sustaining the lives of fellow humans for the simple benefit of being appreciated. Some people contend that this is hardwired in humans. Anyone can get pleasure from feeding others, and ultimately it is not a matter of wishing we would eat together more often. We simply must, for our own happiness.

NOTES

1. Brian Lipinski, Craig Hanson, James Lomax, Lisa Kitinoja, Richard Waite, and Tim Searchinger, "Reducing Food Loss and Waste," working paper, installment 2 of *Creating a Sustainable Food Future* (Washington, D.C.: World Resources Institute, 2013), 5.
2. Janet Flammang, *The Taste for Civilization* (Champaign: University of Illinois Press, 2009).
3. Lindsey Smith, Shu Wen Ng, and Barry M. Popkin, "Trends in U.S. Home Food Preparation and Consumption: Analysis of National Nutrition Surveys and Time Use Studies from 1965–1966 to 2007–2008," *Nutrition Journal* 12, no. 45 (April 2013): 1.
4. Alice Julier, *Eating Together* (Champaign: University of Illinois Press, 2013).

13

A Plea for Culinary Modernism

Why We Should Love Fast, Modern, Processed Food
(With a New Postscript)

RACHEL LAUDAN

Modern, fast, processed food is a disaster. That, at least, is the message conveyed by newspapers and magazines, on television cooking programs, and in prizewinning cookbooks. It is a mark of sophistication to bemoan the steel roller mill and supermarket bread while yearning for stone-ground flour and brick ovens; to seek out heirloom apples and pumpkins while despising modern tomatoes and hybrid corn; to be hostile to agronomists who develop high-yielding modern crops and to home economists who invent new recipes for General Mills. We hover between ridicule and shame when we remember how our mothers and grandmothers enthusiastically embraced canned and frozen foods. We nod in agreement when the waiter proclaims that the restaurant showcases the freshest local produce. We shun Wonder Bread and Coca-Cola. Above all, we loathe the great culminating symbol of Culinary Modernism, McDonald's—modern, fast, homogeneous, and international.

Like so many of my generation, my culinary style was created by those who scorned industrialized food; Culinary Luddites, we may call them, after the English handworkers of the nineteenth century who abhorred the machines that were destroying their traditional way of life. I learned to cook from the books of Elizabeth David, who urged us to sweep our store cupboards "clean forever of the cluttering debris of commercial sauce bottles and all synthetic flavorings." I progressed to the Time-Life *Good Cook* series and to *Simple French Cooking*, in which Richard Olney hoped against hope that "the reins of stubborn habit are strong enough to frustrate the famous industrial revolution for some time to come."[1] I turned to Paula Wolfert to learn more about Mediterranean cooking and was assured that I wouldn't "find a dishonest dish in this book. . . . The food here is real food . . . the real food of real people." Today I rush to the newsstand to pick up *Saveur*, with its promise to teach me to "savor a world of authentic cuisine."

Culinary Luddism involves more than just taste. Since the days of the counterculture, it has also presented itself as a moral and political crusade. Now in Boston, the Oldways Preservation and Exchange Trust works to provide "a scientific basis for the preservation and revitalization of traditional diets."[2] Meanwhile, Slow Food, founded in 1989 to protest the opening of a McDonald's in Rome, is a self-described "Greenpeace for food"; its manifesto begins, "We are enslaved by speed and have all succumbed to the same insidious virus: Fast Life, which disrupts our habits, pervades the privacy of our homes and forces us to eat Fast Foods. . . . Slow Food is now the only truly progressive answer."[3] As one of its spokesmen was reported as saying in the *New York Times*, "Our real enemy is the obtuse consumer."[4]

At this point I begin to back off. I want to cry, "Enough!" But why? Why would I, who learned to cook from Culinary Luddites, who grew up in a family that, in Elizabeth David's words, produced their "own home-cured bacon, ham and sausages . . . churned their own butter, fed their chickens and geese, cherished their fruit trees, skinned and cleaned their own hares" (well, to be honest, not the geese and sausages), not rejoice at the growth of Culinary Luddism?[5] Why would I (or anyone else) want to be thought "an obtuse consumer"? Or admit to preferring unreal food for unreal people? Or to savoring inauthentic cuisine?

The answer is not hard to find: because I am a historian. As a historian I cannot accept the account of the past implied by Culinary Luddism, a past sharply divided between good and bad, between the sunny rural days of yore and the gray industrial present. My enthusiasm for Luddite kitchen wisdom does not carry over to their history, any more than my response to a stirring political speech inclines me to accept the orator as scholar. The Luddites' fable of disaster, of a fall from grace, smacks more of wishful thinking than of digging through archives. It gains credence not from scholarship but from evocative dichotomies: fresh and natural versus processed and preserved; local versus global; slow versus fast; artisanal and traditional versus urban and industrial; healthful versus contaminated and fatty. History shows, I believe, that the Luddites have things back to front.

That food should be fresh and natural has become an article of faith. It comes as something of a shock to realize that this is a latter-day creed. For our ancestors, natural was something quite nasty. Natural often tasted bad. Fresh meat was rank and tough, fresh milk warm and unmistakably a bodily excretion; fresh fruits (dates and grapes being rare exceptions outside the tropics) were inedibly sour, fresh vegetables bitter. Even today, natural can be a shock when we actually encounter it. When Jacques Pepin offered free-range chickens to friends, they found "the flesh tough and the flavor too strong," prompting him to wonder whether they would really like things the way they naturally used to be.[6]

Natural was unreliable. Fresh fish began to stink, fresh milk soured, eggs went rotten. Everywhere seasons of plenty were followed by seasons of hunger when the days were short, the weather turned cold, or the rain did not fall. Hens stopped laying eggs, cows went dry, fruits and vegetables were not to be found, fish could not be caught in the stormy seas. Natural was usually indigestible. Grains, which supplied from 50% to 90% of the calories in most societies, have to be threshed, ground, and cooked to make them edible. Other plants, including the roots and tubers that were the life support of the societies that did not eat grains, are often downright poisonous. Without careful processing, green potatoes, stinging taro, and cassava, bitter with prussic acid, are not just indigestible but toxic.

Nor did our ancestors' physiological theories dispose them to the natural. Until about two hundred years ago, from China to Europe, and in Mesoamerica, too, everyone believed that the fires in the belly cooked foodstuffs and turned them into nutrients.[7] That was what digesting was. Cooking foods in effect predigested them and made them easier to assimilate. Given a choice, no one would burden the stomach with raw, unprocessed foods.

So to make food tasty, safe, digestible, and healthy, our forebears bred, ground, soaked, leached, curdled, fermented, and cooked naturally occurring plants and animals until they were literally beaten into submission. To lower toxin levels, they cooked plants, treated them with clay (the Kaopectate effect), and leached them with water, acid fruits and vinegars, and alkaline lye.[8] They intensively bred maize to the point that it could not reproduce without human help. They created sweet oranges and juicy apples and non-bitter legumes, happily abandoning their more natural but less tasty ancestors. They built granaries for their grain, dried their meat and their fruit, salted and smoked their fish, curdled and fermented their dairy products, and cheerfully used whatever additives and preservatives they could—sugar, salt, oil, vinegar, lye—to make edible foodstuffs. In the twelfth century, the Chinese sage Wu Tzu-mu listed the six foodstuffs essential to life: rice, salt, vinegar, soy sauce, oil, and tea.[9] Four had been unrecognizably transformed from their naturally occurring state. Who could have imagined vinegar as rice that had been fermented to ale and then soured? Or soy sauce as cooked and fermented beans? Or oil as the extract of crushed cabbage seeds? Or bricks of tea as leaves that had been killed by heat, powdered, and compressed? Only salt and rice had any claim to fresh or natural, and even then the latter had been stored for months or years, threshed, and husked.

Processed and preserved foods kept well, were easier to digest, and were delicious: raised white bread instead of chewy wheat porridge; thick, nutritious, heady beer instead of prickly grains of barley; unctuous olive oil instead of a tiny, bitter fruit; soy milk, sauce, and tofu instead of dreary, flatulent soybeans; flexible, fragrant tortillas instead of dry, tough maize; not to mention red wine, blue cheese, sauerkraut, hundred-year-old eggs, Smithfield hams, smoked salmon, yogurt, sugar, chocolate, and fish sauce.

Eating fresh, natural food was regarded with suspicion verging on horror, something to which only the uncivilized, the poor, and the starving resorted.[10] When the compiler of the Confucian classic the *Book of Rites* (ca. 200 B.C.E.), distinguished the first humans—people who had no alternative to wild, uncooked foods—from civilized peoples who took "advantage of the benefits of fire . . . [who] toasted, grilled, boiled, and roasted," he was only repeating a commonplace.[11] When the ancient Greeks took it as a sign of bad times if people were driven to eat greens and root vegetables, they too were rehearsing common wisdom.[12] Happiness was not a verdant Garden of Eden abounding in fresh fruits, but a securely locked storehouse jammed with preserved, processed foods.

Local food was greeted with about as much enthusiasm as fresh and natural. Local foods were the lot of the poor who could neither escape the tyranny of local climate and biology nor the monotonous, often precarious diet it afforded. Meanwhile, the rich, in search of a more varied diet, bought, stole, wheedled, robbed, taxed, and ran off with appealing plants and animals, foodstuffs, and culinary techniques from wherever they could find them.

By the fifth century B.C.E., Celtic princes in the region of France now known as Burgundy were enjoying a glass or two of Greek wine, drunk from silver copies of Greek drinking vessels.[13] The Greeks themselves looked to the Persians, acclimatizing their peaches and apricots and citrons and emulating their rich sauces, while the Romans in turn hired Greek cooks. From around the time of the birth of Christ, the wealthy in China, India, and the Roman Empire paid vast sums for spices brought from the distant and mysterious Spice Islands. From the seventh century C.E., Islamic caliphs and sultans transplanted sugar, rice, citrus, and a host of other Indian and Southeast Asian plants to Persia and the Mediterranean, transforming the diets of West Asia and the shores of the Mediterranean. In the thirteenth century, the Japanese had naturalized the tea plant of China and were importing sugar from Southeast Asia. In the seventeenth century, the European rich drank sweetened coffee, tea, and cocoa in Chinese porcelain, imported or imitation, proffered by servants in Turkish or other foreign dress. To ensure their own supply, the French, Dutch, and English embarked on imperial ventures and moved millions of Africans and Asians around the globe. The Swedes, who had no empire, had a hard time getting these exotic foodstuffs, so the eighteenth-century botanist Linnaeus set afoot plans to naturalize the tea plant in Sweden.

We may laugh at the climatic hopelessness of his proposal. Yet it was no more ridiculous than other, more successful, proposals to naturalize Southeast Asian sugarcane throughout the tropics, apples in Australia, grapes in Chile, Hereford cattle in Colorado and Argentina, and Caucasian wheat on the Canadian prairie.[14] Without our aggressively global ancestors, we would all still be subject to the tyranny of the local.

As for slow food, it is easy to wax nostalgic about a time when families and friends met to relax over delicious food, and to forget that, far from being an invention of the late twentieth century, fast food has been a mainstay of every society. Hunters tracking their prey, fishermen at sea, shepherds tending their flocks, soldiers on campaign, and farmers rushing to get in the harvest all needed food that could be eaten quickly and away from home. The Greeks roasted barley and ground it into a meal to eat straight or mixed with water, milk, or butter (as the Tibetans still do), while the Aztecs ground roasted maize and mixed it with water to make an instant beverage (as the Mexicans still do).[15]

City dwellers, above all, relied on fast food. When fuel costs as much as the food itself, when huddled dwellings lacked cooking facilities, and when cooking fires might easily conflagrate entire neighborhoods, it made sense to purchase your bread or noodles, and a little meat or fish to liven them up. Before the birth of Christ, Romans were picking up honey cakes and sausages in the Forum.[16] In twelfth-century Hangchow, the Chinese downed noodles, stuffed buns, bowls of soup, and deep-fried confections. In Baghdad of the same period, the townspeople bought ready-cooked meats, salt fish, bread, and a broth of dried chickpeas. In the sixteenth century, when the Spanish arrived in Mexico, Mexicans had been enjoying tacos from the market for generations. In the eighteenth century, the French purchased cocoa, apple turnovers, and wine in the boulevards of Paris, while the Japanese savored tea, noodles, and stewed fish.

Deep-fried foods, expensive and dangerous to prepare at home, have always had their place on the street: doughnuts in Europe, churros in Mexico, *andagi* in Okinawa, and *sev* in India. Bread, also expensive to bake at home, is one of the oldest convenience foods. For many people in West Asia and Europe, a loaf fresh from the baker was the only warm food of the day. To these venerable traditions of fast food, Americans have simply added the electric deep fryer, the heavy iron griddle of the Low Countries, and the franchise.[17] The McDonald's in Rome was, in fact, just one more in a long tradition of fast food joints reaching back to the days of the Caesars.

What about the idea that the best food was country food, handmade by artisans?[18] That food came from the country goes without saying. The presumed corollary—that country people ate better than city dwellers—does not. Few who worked the land were independent peasants baking their own bread, brewing their own wine or beer, and salting down their own pig. Most were burdened with heavy taxes and rents paid in kind (that is, food); or worse, they were indentured servants, serfs, or slaves. Barely part of the cash economy, they subsisted on what was left over. "The city dwellers," remarked the great Roman doctor Galen in the second century C.E., "collected and stored enough grain for all the coming year immediately after the harvest. They carried off all the wheat, the barley, the beans and the lentils and left what remained to the countryfolk."[19] What remained was pitiful. All too often, those who worked the land got by on thin gruels and gritty flatbreads. North of the Alps, French peasants prayed that chestnuts would be sufficient to sustain them from the time when their grain ran out to the harvest still three months away.[20] South of the Alps, Italian peasants suffered skin eruptions, went mad, and in the worst cases died of pellagra brought on by a diet of maize (i.e., corn) polenta and water. The dishes we call ethnic and assume to be of peasant origin were invented for the urban, or at least urbane, aristocrats who collected the surplus. This is as true of the lasagna of northern Italy as it is of the chicken korma of Mughal Delhi; the moo shu pork of imperial China; the pilafs, stuffed vegetables, and baklava of the great Ottoman palace in Istanbul; or the *mee krob* of nineteenth-century Bangkok.[21] Cities have always enjoyed the best food and have invariably been the focal points of culinary innovation.

Nor are most "traditional foods" very old. For every prized dish that goes back two thousand years, a dozen have been invented in the last two hundred.[22] The French baguette? A twentieth-century phenomenon, adopted nationwide only after World War II. English fish and chips? Dates from the late nineteenth century, when the working class took up the fried fish of Sephardic Jewish immigrants in East London. Fish and chips, though, will soon be a thing of the past. It's a Balti and lager now, Balti being a kind of stir-fried curry dreamed up by Pakistanis living in Birmingham. Greek moussaka? Created in the early twentieth century in an attempt to Frenchify Greek food. The bubbling Russian samovar? Late eighteenth century. The Indonesian rijsttafel? Dutch colonial food. Indonesian *padang* food? Invented for the tourist market in the past fifty years. Tequila? Promoted as the national drink of Mexico during the 1930s by the Mexican film industry. Indian tandoori chicken? The brainchild of Hindu Punjabis who survived by selling chicken cooked in a Muslim-style tandoor oven when they fled Pakistan for Delhi during the Partition of India. The soy sauce, steamed white rice, sushi, and tempura of Japan? Commonly eaten only after the middle of the nineteenth century. The *lomilomi* salmon, salted salmon rubbed with chopped tomatoes and spring onions, that is a fixture in every Hawaiian luau? Not a salmon is to be found within two thousand miles of the islands, and onions and tomatoes were unknown in Hawaii until the nineteenth century. These are indisputable facts of history, though if you point them out you will be met with stares of disbelief.

Not only were many "traditional" foods created after industrialization and urbanization, a lot of them were dependent on it. The Swedish smorgasbord came into its own at the beginning of the twentieth century, when canned out-of-season fish, roe, and liver paste made it possible to set out a lavish table. Hungarian goulash was unknown before the nineteenth century, and not widely accepted until after the invention of a paprika-grinding mill in 1859.[23]

When lands were conquered, peoples migrated, populations converted to different religions or accepted new dietary theories, and dishes—even whole cuisines—were forgotten and new ones invented. Where now is the cuisine of Renaissance Spain and Italy, or of the Indian Raj, or of Tsarist Russia, or of medieval Japan? Instead we have Nyonya food in Singapore, Cape Malay food in South Africa, Creole food in the Mississippi Delta, and "local food" in Hawaii. How long does it take to create a cuisine? Not long: less than fifty years, judging by past experience.

Were old foods more healthful than ours? Inherent in this vague notion are several different claims, among them that foods were less dangerous, that diets were better balanced. Yet while we fret about pesticides on apples, mercury in tuna, and mad cow disease, we should remember that ingesting food is, and always has been, inherently dangerous. Many plants contain both toxins and carcinogens, often at levels much higher than any pesticide residues.[24] Grilling and frying add more. Some historians argue that bread made from moldy, verminous flour, or adulterated with mash, leaves, or bark to make it go further, or contaminated with hemp or poppy seeds to drown out sorrows, meant that for five hundred years Europe's poor staggered around in a drugged haze subject to hallucinations.[25] Certainly, many of our forebears were drunk much of the time, given that beer or wine were preferred to water, and with good reason. In the cities, polluted water supplies brought intestinal diseases in their wake. In France, for example, no piped water was available until the 1860s. Bread was likely to be stretched with chalk, pepper adulterated with the sweepings of warehouse floors, and sausage stuffed with all the horrors famously exposed by Upton Sinclair in *The Jungle*. Even the most reputable cookbooks recommended using concentrated sulphuric acid to intensify the color of jams.[26] Milk, suspected of spreading scarlet fever, typhoid, and diphtheria as well as tuberculosis, was sensibly avoided well into the twentieth century when the United States and many parts of Europe introduced stringent regulations. My mother sifted weevils from the flour bin; my aunt reckoned that if the maggots could eat her home-cured ham and survive, so could the family.

As to dietary balance, once again we have to distinguish between rich and poor. The rich, whose bountiful tables and ample girths were visible evidence of their station in life, suffered many of the diseases of excess. In the seventeenth century, the Mughal emperor Jahangir died of overindulgence in food, opium, and alcohol.[27] In Georgian England, George Cheyne, the leading doctor, had to be wedged in and out of his carriage by his servants when he soared to four hundred pounds, while a little later Erasmus Darwin, grandfather of Charles and another important physician, had a semicircle cut out of his dining table to accommodate his paunch. In the nineteenth century, the fourteenth shogun of Japan died at age twenty-one, probably of beriberi induced by eating the white rice available only to the privileged. In the Islamic countries, India, and Europe, the well-to-do took sugar as a medicine; in India they used butter; and in much of the world people avoided fresh vegetables, all on medical advice.

Whether the peasants really starved, and if so how often, particularly outside Europe, is the subject of ongoing research.[28] What is clear is that the food supply was always precarious: if the weather was bad or war broke out, there might not be enough to go around. The end of winter or the dry season saw everyone suffering from the lack of fresh fruits and vegetables, scurvy occurring on land as well as at sea. By our standards, the diet was scanty for people who were engaged in heavy physical toil. Estimates suggest that in France on the eve of the Revolution, one in three adult men got by on no more than 1,800 calories a day, while a century later in Japan daily intake was perhaps 1,850 calories. Historians believe that in times of scarcity, peasants essentially hibernated during the winter.[29] It is not surprising, therefore, that in France the proudest of boasts was "There is always bread in the house," while the Japanese adage advised, "All that matters is a full stomach."[30]

By the standard measures of health and nutrition—life expectancy and height—our ancestors were far worse off than we are.[31] Much of the blame was due to the diet, exacerbated by living conditions and infections which affect the body's ability to use the food that is ingested.[32] No amount of nostalgia for the pastoral foods of the distant past can wish away the fact that our ancestors lived mean, short lives, constantly afflicted with diseases, many of which can be directly attributed to what they did and did not eat.

Historical myths, though, can mislead as much by what they don't say as by what they do. Culinary Luddites typically gloss over the moral problems intrinsic to the labor of producing and preparing food. In 1800, 95% of the Russian population and 80% of the French lived in the country; in other words, they spent their days getting food on the table for themselves and other people. A century later, 88% of Russians, 85% of Greeks, and over 50% of the French were still on the land.[33] Traditional societies were aristocratic, made up of the many who toiled to produce, process, preserve, and prepare food, and the few who, supported by the limited surplus, could do other things.

In the great kitchens of the few—royalty, aristocracy, and rich merchants—cooks created elaborate cuisines. The cuisines drove home the power of the mighty few with a symbol that everyone understood: ostentatious shows of more food than the powerful could possibly consume. Feasts were public occasions for the display of power, not private occasions for celebration, for enjoying food for food's sake. The poor were invited to watch, groveling as the rich gorged themselves.[34] Louis XIV was exploiting a tradition going back to the Roman Empire when he encouraged spectators at his feasts. Sometimes, to hammer home the point while amusing the court, the spectators were let loose on the leftovers. "The destruction of so handsome an arrangement served to give another agreeable entertainment to the court," observed a commentator, "by the alacrity and disorder of those who demolished these castles of marzipan, and these mountains of preserved fruit."[35]

Meanwhile, most men were born to a life of labor in the fields, most women to a life of grinding, chopping, and cooking. "Servitude," said my mother as she prepared home-cooked breakfast, dinner, and tea for eight to ten people 365 days a year. She was right. Churning butter and skinning and cleaning hares, without the option of picking up the phone for a pizza if something goes wrong, is unremitting, unforgiving toil. Perhaps, though, my mother did not realize how much worse her lot might have been. She could at least buy our bread from the bakery. In Mexico, at the same time, women without servants could expect to spend five hours a day—one-third of their waking hours—kneeling at the grindstone preparing the dough for the family's tortillas. Not until the 1950s did the invention of the tortilla machine release them from the drudgery.[36]

In the eighteenth and early nineteenth centuries, it looked as if the distinction between gorgers and grovelers would worsen. Between 1575 and 1825 world population had doubled from 500 million to a billion, and it was to double again by 1925. Malthus sounded his dire predictions. The poor, driven by necessity or government mandate, resorted to basic foods that produced bountifully even if they were disliked: maize and sweet potatoes in China and Japan, maize in Italy, Spain, and Romania, potatoes in northern Europe.[37] They eked out an existence on porridges or polentas of oats or maize, on coarse breads of rye or barley bulked out with chaff or even clay and ground bark, and on boiled potatoes; they saw meat only on rare occasions.[38] The privation continued. In Europe, 1845 was a year of hunger, best remembered now as the beginning of the dire potato famine of Ireland. Meanwhile, the rich continued to indulge, feasting on white bread, meats, rich fatty sauces, sweet desserts, exotic hothouse-grown pineapples, wine, and tea, coffee, and chocolate drunk from fine china. That same year, just before revolutions rocked Europe, the British prime minister Benjamin Disraeli described "two nations, between whom there is no intercourse and no sympathy . . . who are formed by a different breeding, are fed by a different food, are ordered by different manners, and are not governed by the same laws . . . THE RICH AND THE POOR."[39]

In the nick of time, in the 1880s, the industrialization of food got under way long after the production of other common items of consumption such as textiles and clothing had been mechanized. Farmers brought new land into production, utilized reapers and later tractors and combines, spread more fertilizer, and, by the 1930s, began growing hybrid maize. Steamships and trains brought fresh and canned meats, fruits, vegetables, and milk to the growing towns. Instead of starving, the poor of the industrialized world survived and thrived. In Britain the retail price of food in a typical workman's budget fell by a third between 1877 and 1887 (though he would still spend 71% of his income on food and drink). In 1898 in the United States a dollar bought 42% more milk, 51% more coffee, a third more beef, twice as much sugar, and twice as much flour as in 1872.[40] By the beginning of the twentieth century, members of the British working class were drinking sugary tea from china teacups and eating white bread spread with jam and margarine, canned meats, canned pineapple, and an orange from the Christmas stocking.

To us, the cheap jam, the margarine, and the starchy diet look pathetic. Yet white bread did not cause the "weakness, indigestion, or nausea" that coarse whole wheat bread did when it supplied most of the calories (not a problem for us since we never consume it in such quantities).[41] Besides, it was easier to detect stretchers such as sawdust in white bread. Margarine and jam made the bread more attractive and easier to swallow. Sugar tasted good, and hot tea in an unheated house in midwinter provided good cheer. For those for whom fruit had been available, if at all, only from June to October, canned pineapple and a Christmas orange were treats to be relished. For the diners, therefore, the meals were a dream come true, a first step away from a coarse, monotonous diet and the constant threat of hunger, even starvation.

Nor should we think it was only the British, not famed for their cuisine, who were delighted with industrialized foods. Everyone was, whether American, Asian, African, or European. In the first half of the twentieth century, Italians embraced factory-made pasta and canned tomatoes.[42] In the second half of the century, Japanese women welcomed factory-made bread because they could sleep in a little longer instead of having to get up to make rice.[43] Similarly, Mexicans seized on bread as a good food to have on hand when there was no time to prepare tortillas. Working women in India are happy to serve commercially made bread during the week, saving the time-consuming business of making chapatis for the weekend. As supermarkets appeared in Eastern Europe and Russia, housewives rejoiced at the choice and convenience of ready-made goods. For all, Culinary Modernism had provided what was wanted: food that was processed, preservable, industrial, novel, and fast, the food of the elite at a price everyone could afford. Where modern food became available, populations grew taller, stronger, had fewer diseases, and lived longer. Men had choices other than hard agricultural labor, women other than kneeling at the metate five hours a day.

So the sunlit past of the Culinary Luddites never existed. So their ethos is based not on history but on a fairy tale. So what? Perhaps we now need this culinary philosophy. Certainly no one would deny that an industrialized food supply has its own problems—problems we hear about every day. Perhaps we should eat more fresh, natural, local, artisanal, slow food. Why not create a historical myth to further that end? The past is over and gone. Does it matter if the history is not quite right?

It matters quite a bit, I believe. If we do not understand that most people had no choice but to devote their lives to growing and cooking food, we are incapable of comprehending that the foods of Culinary Modernism—egalitarian, available more or less equally to all, without demanding the disproportionate amount of the resources of time or money that traditional foodstuffs did—allow us unparalleled choices not just of diet but of what to do with our lives. If we urge the Mexican to stay at her metate, the farmer to stay at his olive press, the housewife to stay at her stove instead of going to McDonald's, all so that we may eat handmade tortillas, traditionally pressed olive oil, and home-cooked meals, we are assuming the mantle of the aristocrats of old. We are reducing the options of others as we attempt to impose our elite culinary preferences on the rest of the population.

If we fail to understand how scant and monotonous most traditional diets were, we can misunderstand the "ethnic foods" we encounter in cookbooks, restaurants, or on our travels. We let our eyes glide over the occasional references to servants, to travel and education abroad in so-called ethnic cookbooks, references that otherwise would clue us in to the fact that the recipes are those of monied Italians, Indians, or Chinese with maids to do the donkey work of preparing elaborate dishes. We may mistake the meals of today's European, Asian, or Mexican middle class (many of them benefiting from industrialization and contemporary tourism) for peasant food or for the daily fare of our ancestors. We can represent the peoples of the Mediterranean, Southeast Asia, India, or Mexico as pawns at the mercy of multinational corporations bent on selling trashy modern products—failing to appreciate that, like us, they enjoy a choice of goods in the market, foreign restaurants to eat at, and new recipes to try. A Mexican friend, suffering from one too many foreign visitors who chided her because she offered Italian rather than Mexican food, complained, "Why can't we eat spaghetti, too?"

If we unthinkingly assume that good food maps neatly onto old or slow or homemade food (even though we've all had lousy traditional cooking), we miss the fact that lots of industrial foodstuffs are better. Certainly no one with a grindstone will ever produce chocolate as suave as that produced by conching in a machine for seventy-two hours. Nor is the housewife likely to turn out fine soy sauce or miso. And let us not forget that the current popularity of Italian food owes much to the availability and long shelf life of two convenience foods that even purists love, high-quality factory pasta and canned tomatoes. Far from fleeing them, we should be clamoring for more high-quality industrial foods.

If we romanticize the past, we may miss the fact that it is the modern, global, industrial economy (not the local resources of the wintry country around New York, Boston, or Chicago) that allows us to savor traditional, peasant, fresh, and natural foods. Virgin olive oil, Thai fish sauce, and udon noodles come to us thanks to international marketing. Fresh and natural loom so large because we can take for granted the preserved and processed staples—salt, flour, sugar, chocolate, oils, coffee, tea—produced by agribusiness and food corporations. Asparagus and strawberries in winter come to us on trucks trundling up from Mexico and planes flying in from Chile. Visits to charming little restaurants and colorful markets in Morocco or Vietnam would be impossible without international tourism. The ethnic foods we seek out when we travel are being preserved, indeed often created, by a hotel and restaurant industry determined to cater to our dream of India or Indonesia, Turkey, Hawaii, or Mexico.[44] Culinary Luddism, far from escaping the modern global food economy, is parasitic upon it.

Culinary Luddites are right, though, about two important things. We need to know how to prepare good food, and we need a culinary ethos. As far as good food goes, they've done us all a service by teaching us how to use the bounty delivered to us (ironically) by the global economy. Their culinary ethos, though, is another matter. Were we able to turn back the clock, as they urge, most of us would be toiling all day in the fields or the kitchen; many of us would be starving. Nostalgia is not what we need. What we need is an ethos that comes to terms with contemporary, industrialized food, not one that dismisses it, an ethos that opens choices for everyone, not one that closes them for many so that a few may enjoy their labor, and an ethos that does not prejudge, but decides case by case when natural is preferable to processed, fresh to preserved, old to new, slow to fast, artisanal to industrial. Such an ethos, and not a timorous Luddism, is what will impel us to create the matchless modern cuisines appropriate to our time.

POSTSCRIPT

"A Plea for Culinary Modernism" was published in 2001, over fifteen years before the other essays for this volume were written.[45] Since then, much has changed in American food politics. Those unhappy with contemporary food shifted the focus of their attacks from McDonald's first to Walmart and more recently to Monsanto. Gluten-free and low-carb have joined fresh, organic, and natural as desirable. "Celebrity chef," "sustainability," and "locavore" have become household words. Fats and sweeteners have been vilified and unvilified. In universities, food studies courses and programs, many of them laying out an activist agenda, have multiplied.

This postscript is thus the perfect opportunity to pick up on the observation shared by Margot Finn and Steve Striffler: namely, that after two decades, many Americans, perhaps the majority, although confused about what to eat, have not responded to the repeated calls to change modern food. This, I believe, is because beneath the back-and-forth about the "facts of the matter" lie deep differences about what I call culinary philosophy. I use the term "philosophy" because our ideas about what is just, fair, healthful for the body and the earth, and tasty have always shaped what we eat. (To prevent confusion, readers should note that this a broader notion of the philosophy relevant to food than Robert Valgenti's.) I prefix "philosophy" with the word "culinary" (relating to the kitchen or cookery) because in the process of food preparation or cookery we transform what nature provides into what our philosophy prescribes. If everyone shared the same culinary philosophy, debates about food would be reduced to matters of fact. People don't, however, and never have. So to make explicit the underlying culinary philosophies, I will sketch the four that are most important to understanding the contemporary situation: monarchical; republican; romantic; and socialist.

Monarchical or aristocratic culinary philosophy was the norm among the European aristocracy in the mid-eighteenth century (and indeed around the world through much of history).[46] According to this, the ruling class was entitled to a more refined and luxurious cuisine than its subjects. At aristocratic dinners or (later) high-end restaurants, men, sometimes accompanied by women, dined off fine china while conversing about affairs of state or high culture as waiters hovered nearby. The refined food (we would say highly processed) consisted of richly sauced meats with fine white bread and delicate desserts, all prepared by professional male cooks. These dishes stimulated their appetites, were readily digestible by their delicate constitutions, and demonstrated their rank. Drinks, as Charles Ludington shows, paralleled foods in expressing and reinforcing social status. In Great Britain, for example, fine clarets were a sign of both high class and good taste, while beer was for ordinary people, and port for those who aspired to move up the social scale. Meanwhile, the workers who labored on great estates and in large kitchens to provide and prepare the food were thought to be able to get by with much rougher, simpler fare such as dark bread and root vegetables, readily digestible by their coarse constitutions.

Americans, with important exceptions such as southern planters, turned their backs on monarchical culinary philosophy, opting instead for republican culinary philosophy. Also with roots going back to Antiquity, but coming to the fore in the eighteenth century, this held that all citizens should enjoy simple, bountiful food produced on family farms by free yeoman farmers. Taste did not differ according to rank argued David Hume and Immanuel Kant (and as Valgenti shows), at least if diners applied themselves to learning discrimination. At the meal shared by the whole family, children absorbed the physical nourishment that made them strong citizens. Moderate portions of simple, home-cooked, satisfying foods such as bread, meat, beans and cabbage, and cheese and eggs satisfied healthy appetites and made them strong and vigorous. At the same time, conversing with their parents, children imbibed the mental and moral nourishment necessary to the education of virtuous citizens, coming to understand that they should be courageous, straightforward, dignified, civil, and obedient to the call of duty.[47] When Abraham Lincoln declared Thanksgiving a national holiday in 1863, the ample, simple, inclusive family meal was enshrined as a national ideal.

Republican culinary philosophy continued to underpin food policy over the next century as the nation struggled to include immigrants, women, and African Americans as citizens, to feed a growing population, to deal with migration from country to city, and to fight wars at home and overseas. It provided the underlying framework for the domestic science, home economics, and National Nutrition Program movements described by Charlotte Biltekoff, the efforts to provide safe, nutritious food for babies that Amy Bentley discusses, and the government encouragement of homesteading, subsidies of irrigation water, agricultural research and outreach, or support for family farms laid out by Peter Coclanis and Sarah Ludington.

By the 1970s, though, tensions within mainstream American republican culinary philosophy could no longer be ignored.[48] If most farms were still in family hands, they were now farmed using giant machinery, fossil fuels, and chemicals to aid growth and control pests. Margaret Mellon and Steve Striffler, like many Americans, are cautious about, often deeply opposed to, industrial agriculture. Environment threats, GMOs, and the use of low-paid immigrant labor in the fields were all causes of worry. If the free market remained an ideal, government subsidies and large corporations made it unattainable. If the family meal was still the aspiration of many, it could also be a cause of tension, as Ken Albala reminds readers. As Tracey Deutsch emphasizes (a point with which I heartily agree), preparing the meal was a burden that fell unequally on women, particularly servants and working women, and besides, many ate in canteens or cafés, not around the family table. Well-meaning efforts to improve diets of immigrants and the working class, as Biltekoff makes clear, expressed and mirrored social concern, so that if these groups failed to eat the food regarded as good by the middle classes, it was all too easily assumed that they could not be good people.

So what was to be done? Returning to monarchical culinary philosophy was unthinkable in America, so many opted for a different alternative—romantic culinary philosophy. The restaurateur Alice Waters, proprietor of Chez Panisse in San Francisco; the writer and political activist Carlo Petrini, who founded the Slow Food movement; and Dun Gifford, founder of Oldways, were prominent among those arguing that Americans (and others around the world) would eat more enjoyably and more healthfully if they returned to old ways and traditional peasant diets.[49] Knowingly or not, they were following in the footsteps of the eighteenth-century French philosopher Jean-Jacques Rousseau. In *Émile*, his 1762 treatise on education, Rousseau inveighed against monarchical cuisines with hovering flunkeys serving sauces and sweets to gluttonous, overfed aristocrats. Yet even though he shared with republican culinary philosophy the distaste for the aristocratic meal and the beliefs that underpinned it, Rousseau had no enthusiasm for the family meal, preferring a picnic of freshly gathered and lightly cooked fruits and vegetables taken outdoors in the natural world with friends who reflected on nature and natural food.

In the same vein, after surveying American food options in his best-selling book *The Omnivore's Dilemma* (2006), Michael Pollan concluded with his perfect meal: a meal of foods he had acquired himself (many by foraging) and shared with friends. This was an example of the slow food that diners should embrace in preference to fast food, particularly the McDonald's hamburger, which as it "McDonaldized" the world exemplified the overreach of American capitalism and imperialism. Cooking should be seen neither as a servile task demanded by aristocrats nor as a dutiful responsibility to the family. Instead it was a creative endeavor, in which, as Ken Albala's paean to cooking intimates, the cook was on a par with the poet or artist in guiding his audience into an appreciation of the natural world. Eating well was not just a matter of affirming status or maintaining health but could in and of itself produce political change in farming, the food economy, and society. This optimism was neatly summed by the food journalist Corby Kummer in the title of an enthusiastic article on Slow Food, "Doing Well by Eating Well."[50]

Yet could shopping, cooking, and eating really bring about meaningful social change? Were those who sought out the perfect peach not slipping back into something scarily reminiscent of monarchical culinary philosophy, satisfied with a farmers market and a Whole Foods in town, and seeing no need for further change? With no larger social or economic framework, romantic culinary philosophy lost its larger relevance, constantly subject to the charge, which Margot Finn and Steve Striffler endorse, that it was elitist, that only the rich could eat like peasants (or "peasants").

Socialist culinary philosophy, the final major alternative, did offer a political agenda. Like the other three culinary philosophies, it had roots going back to the Enlightenment and beyond. In its most general formulation, it was understood to mean that food should be produced, prepared, and consumed communally and that the perfect meal was the communal meal. Throughout the nineteenth century, utopian socialists in Europe and religious communities in the United States experimented with various ways of organizing communal farming, cooking, and eating.[51] In the twentieth century, enormous states, particularly the Soviet Union and Maoist China, tried to create socialist utopias, with often calamitous results.

In the 1970s, the "New Left" in America and Britain, now alert to the purges and famines of the Stalin era, reformulated traditional socialism to a smaller-scale, environmentally aware social activism, one wing of which was the "food movement."[52] This gave a prominent role to peasants, making it easy for those concerned about food to slip back and forth between socialist and romantic culinary philosophy. Before founding the Slow Food movement, for example, Carlo Petrini worked with an Italian communist movement, while the journalist Michael Pollan, writing primarily in the romantic vein, also adopts the language of the "food movement." Where romantics saw peasants through the lens of a pastoral arcadianism as happily working in nature to produce simple, delicious food, those who adhered to a socialist culinary philosophy by contrast saw those who did manual work on the land, whether as peasants or migrant workers, and food consumers as being the two exploited poles of a "food system" in which profits were gobbled up by big agriculture and big food. Those in the food movement put their efforts into improving the lot of the workers and increasing the consumer awareness about the shortcomings of the system. During the past two decades, they have urged a "food revolution" to overthrow "our broken food system," that is, the system created by those who adhered to republican culinary philosophy. Thus the food movement seeks, at least in its more radical moments, as Steve Striffler explains, to use food to take on the larger problem of capitalism that according to this view is inherently corrupted by the profit motive.

This postscript is obviously nothing more than a sketch and as such is overly simplistic in many ways. The four culinary philosophies laid out here are not frozen in time; they evolve to deal with new circumstances. Few people subscribe to any one in its entirety, and fewer yet do so knowingly. The point of this exercise is thus not to box people in to positions they do not hold. It is, rather, to alert readers to the fact that very big issues of social and political principle are at stake in what may seem simple matters of fact about taste, GMOs, or farm subsidies.

With hindsight, I realize that my chief target in "A Plea for Culinary Modernism" back in 2001 was the romantic culinary philosophy that dominated the food debates at the beginning of the twenty-first century. I wanted to remind readers that the United States (and the Western world) had pulled off a near-miracle in providing decent food for an increasing proportion of a rapidly growing population. I wanted to point out how health and safety had improved, including the decline in gastrointestinal illnesses described by Matthew Booker, the wiping out of deficiency diseases such as pellagra and scurvy, and the turning of the tide on hunger and starvation. As a woman, I was personally grateful to have opportunities denied to my mother, whose kitchen work was unending.

Today I would happily extend those earlier arguments to those who want a food revolution to replace our capitalist system. However appealing I find socialist culinary philosophy in theory, faced with its unhappy track record in the Soviet Union, Maoist China, Cuba, and Venezuela, not to mention the catastrophe that is North Korea, I am very wary of yet one more attempt to put it into practice.

This is not to say that I believe that the way we currently produce, prepare, and consume food is perfect. There is some truth, sometimes a lot of truth, in many of the criticisms leveled by adherents of romantic or socialist culinary philosophies. But those of us in the mainstream are not blind to the persistence of old problems, such as inequality and hunger, nor to the appearance of new ones, such as diseases of abundance and threats to the environment. I prefer to throw my lot in with those who believe that a twin program of, on the one hand, continuing to modernize and improve the long-standing food system and, on the other, rethinking republican culinary philosophy to open it to, say, greater enjoyment of food, more equity between men and women, and more sensitivity to the environment is the best way forward. So apparently do the large majority of Americans, who have proved resistant to calls to a food revolution.

NOTES

1. Elizabeth David, *French Country Cooking* (1951; reprint, London: Penguin, 1963), 24–25; Richard Olney, *Simple French Food* (1974; reprint, London: Penguin, 1983), 3; Paula Wolfert, *The Cooking of the Eastern Mediterranean: 215 Healthy, Vibrant and Inspired Recipes* (New York: HarperCollins, 1994), 2. Even social historians join in the song: Georges Duby asserts that medieval food "responds to our fierce, gnawing urge to flee the anemic, the bland, fast food, ketchup, and to set sail for new shores." Foreword to Odile Redon, Françoise Sabban, and Silvano Serventi, *The Medieval Kitchen: Recipes from France and Italy*, trans. Edward Schneider (Chicago: University of Chicago Press, 1998), ix.

2. Oldways Preservation and Exchange Trust, http://www.oldwayspt.org/html/meet.htm, accessed 1999.

3. Slow Food, http://www.slow-food.com/principles/manifest.html, accessed 1999.

4. Slow Food, http://www.slow-food.com/principles/press.html, accessed 1999.

5. David, *French Country Cooking*, 10.

6. Julia Child and Jacques Pepin, *Julia and Jacques Cooking at Home* (New York: Alfred A. Knopf, 1999), 263.

7. Rachel Laudan, "Birth of the Modern Diet," *Scientific American*, August 2000; Laudan, "A Kind of Chemistry," *Petits Propos Culinaires* 62 (1999): 8–22.

8. For these toxins and how humans learned to deal with them, see Timothy Johns, *With Bitter Herbs They Shall Eat It: Chemical Ecology and the Origins of Human Diet and Medicine* (Tucson: University of Arizona Press, 1992).

9. Michael Freeman, "Sung," in K. C. Chang, *Food in Chinese Culture: Anthropological and Historical Perspectives* (New Haven, Conn.: Yale University Press, 1977), 151.

10. For the survival of this view in eighteenth-century America, see Trudy Eden, "The Art of Preserving: How Cooks in Colonial Virginia Imitated Nature to Control It," in "The Cultural Topography of Food," ed. Beatrice Fink, special issue, *Eighteenth-Century Life* 23 (1999): 13–23.

11. E. N. Anderson, *The Food of China* (New Haven, Conn.: Yale University Press, 1988), 41–42.

12. Andrew Dalby, *Siren Feasts: A History of Food and Gastronomy in Greece* (London: Routledge), 24–25.

13. For Greek wines in Germany, see T. G. E. Powell, *The Celts* (1958; reprint, London: Thames and Hudson, 1986), 108–14; for Greek emulation of Persian dining habits and acclimatization of fruits, see Andrew Dalby, "Alexander's Culinary Legacy," in *Cooks and Other People: Proceedings of the Oxford Symposium on Food and Cookery, 1995*, ed. Harlan Walker (Totnes, U.K.: Prospect Books, 1996), 81–85, 89; for the spice trade, J. Innes Miller, *The Spice Trade of the Roman Empire* (Oxford: Clarendon, 1969); for the Islamic agricultural revolution, Andrew M. Watson, *Agricultural Innovation in the Early Islamic World* (Cambridge: Cambridge University Press, 1994); for stimulant drinks in Europe, James Walvin, *Fruits of Empire: Exotic Produce and British Taste, 1660–1800* (Cambridge: Cambridge University Press, 1997). For Linnaeus's efforts to naturalize tea and other plants, see Lisbet Koerner, "Nature and Nation in Linnean Travel" (PhD diss., Harvard University, 1994).

14. For the history of sugar, see Philip Curtin, *The Rise and Fall of the Plantation Complex* (Cambridge: Cambridge University Press, 1990), and Sidney Mintz, *Sweetness and Power: The Place of Sugar in Modern History* (New York: Penguin, 1986); for apples in Australia,

Michael Symons, *One Continuous Picnic: A History of Eating in Australia* (Adelaide: Duck Press, 1982), 96–97; for grapes in Chile and the Mediterranean, Tim Unwin, *Wine and the Vine: An Historical Geography of Viticulture and the Wine Trade* (1991; reprint, London: Routledge, 1996), chap. 9.

15. K. D. White, "Farming and Animal Husbandry," in *Civilization of the Ancient Mediterranean: Greece and Rome*, vol. 1, ed. Michael Grant and Rachel Kitzinger (New York: Charles Scribner's, 1988), 236; Rinjing Dorje, *Food in Tibetan Life* (London: Prospect Books, 1985), 61–65; Inga Clendinnen, *Aztecs: An Interpretation* (Cambridge: Cambridge University Press, 1991), 119.

16. For fast food and takeout stands in ancient Rome, see Florence Dupont, *Daily Life in Ancient Rome* (Oxford: Blackwell, 1992), 181; for Hangchow, Chang, *Food in Chinese Culture*, 158–63; for Baghdad, M. A. J. Beg, "A Study of the Cost of Living and Economic Status of Artisans in Abbasid Iraq," *Islamic Quarterly* 16 (1972): 164, and G. Le Strange, *Baghdad during the Abbasid Caliphate from Contemporary Arabic and Persian Sources* (1900; reprint, London: Oxford University Press, 1924), 81–82; for Paris and Edo, Robert M. Isherwood, "The Festivity of the Parisian Boulevards," and James McClain, "Edobashi: Power, Space, and Popular Culture in Edo," both in *Edo and Paris: Urban Life and the State in the Early Modern Era*, ed. James L. McLain, John M. Merrieman, and Ugawa Kaoru (Ithaca, N.Y.: Cornell University Press, 1994), 114, 293–95.

17. Richard Pillsbury, *No Foreign Food: The American Diet in Time and Place* (Boulder, Colo.: Westview, 1998), 175.

18. "Great cuisines have arisen from peasant societies": Symons, *One Continuous Picnic*, 12. Symons is a restaurateur and historian of Australian food.

19. Quoted by Peter Brown, *The World of Late Antiquity* (London: Thames and Hudson, 1971), 12.

20. Daphne Roe, *A Plague of Corn: The Social History of Pellagra* (Ithaca, N.Y.: Cornell University Press, 1973), chap. 5.

21. Lynne Rossetto Kasper, *The Splendid Table: Recipes from Emilia-Romagna, the Heartland of Northern Italian Food* (New York: Morrow, 1994), 165–69; K. T. Achaya, *Indian Food: A Historical Companion* (Delhi: Oxford University Press, 1994), 158–59; Semahat Arsel, project director, *Timeless Tastes: Turkish Culinary Culture* (Istanbul: Divan, 1996), 48–49.

22. For the baguette, see Philip and Mary Hyman, "France," in *The Oxford Companion to Food*, ed. Alan Davidson (Oxford: Oxford University Press, 1999); for moussaka, Aglaia Kremezi, "Nikolas Tselementes," in Walker, *Cooks and Other People*, 167; for fish and chips, John K. Walton, *Fish and Chips and the British Working Class, 1870–1940* (Leicester: Leicester University Press, 1992); for the samovar, Robert Smith, "Whence the Samovar?," *Petits Propos Culinaires* 4 (1980): 57–82; for rijsttafel, Sri Owen, *Indonesian Regional Cooking* (London: St. Martin's, 1994), 22; for *padang* restaurants, Lisa Klopfer, "Padang Restaurants: Creating 'Ethnic' Food in Indonesia," *Food and Foodways* 5 (1993); for tequila, José María Muría, "El agave histórico: Momentos del tequila," *El Tequila: Arte Tradicional de México, Artes de Mexico* 27 (1995): 17–28; for tandoori chicken, Madhur Jaffrey, *An Invitation to Indian Cooking* (New York: Knopf, 1973), 129–30; for soy sauce, sushi, and soba noodles, Susan B. Hanley, *Everyday Things in Premodern Japan: The Hidden Legacy of Material Culture* (Berkeley: University of California Press, 1999), 161.

23. Dale Brown, *The Cooking of Scandinavia* (New York: Time-Life Books, 1968), 93; Louis Szathmary, "Goulash," in Davidson, *The Oxford Companion to Food;* and George Lang, *The Cuisine of Hungary* (1971; reprint, New York: Bonanza, 1990), 134–35.

24. For the natural carcinogens in plants, see Bruce N. Ames and Lois Swirsky Gold, "Environmental Pollution and Cancer: Some Misconceptions," in *Phantom Risk: Scientific Interference and the Law,* ed. Kenneth R. Foster, David E. Bernstein, and Peter W. Huber (Cambridge, Mass.: MIT Press, 1993), 157–60. For toxins, see Johns, *With Bitter Herbs,* chaps. 3–4.

25. Piero Camporesi, *Bread of Dreams: Food and Fantasy in Early Modern Europe,* trans. David Gentilcore (Chicago: University of Chicago Press, 1991), esp. chaps. 12–15; Lynn Martin, *Alcohol, Sex and Gender in Later Medieval and Early Modern Europe* (New York: St. Martin's, 2000); Jean-Pierre Goubert, *The Conquest of Water,* trans. Andrew Wilson (Princeton, N.J.: Princeton University Press, 1989), 58; J. G. Drummond and Anne Wilbraham, *The Englishman's Food: Five Centuries of English Diet* (1939; reprint, London: Pimlico, 1991), chap. 17; Richard Hooker, *A History of Food and Drink in America* (New York: Bobbs-Merrill, 1981), 298–301.

26. Classic Russian cooking: Elena Molokhovets, *A Gift to Young Housewives,* trans. and ed. Joyce Toomre (Bloomington: Indiana University Press, 1992), 107; Daniel Block, "Purity, Economy, and Social Welfare in the Progressive Era Pure Milk Movement," *Journal for the Study of Food and Society* 3 (1999): 22.

27. Roy Porter, "Consumption: Disease of the Consumer Society," in *Consumption and the World of Goods,* ed. John Brewer and Roy Porter (London: Routledge, 1993), 62; Anita Guerrini, *Obesity and Depression in the Enlightenment: The Life and Times of George Cheyne* (Norman: University of Oklahoma Press, 2000); Hanley, *Everyday Things,* 159–60.

28. Peter Laslett, *The World We Have Lost,* 2nd ed. (New York: Charles Scribner's, 1965).

29. Emmanuel Le Roy Ladurie, *Histoire de la France rurale, II* (Paris, 1975), 438–40; Hanley, *Everyday Things,* 91; Peter Stearns, *European Society in Upheaval: Social History since 1750* (1967; reprint, New York: Macmillan, 1975), 18.

30. Olwen Hufton, "Social Conflict and the Grain Supply in Eighteenth-Century France," in *Hunger and History: The Impact of Changing Food Production and Consumption Patterns on Society,* ed. Robert I. Rotberg and Theodore K. Rabb (Cambridge: Cambridge University Press, 1983), 105–33, 133; Hanley, *Everyday Things,* 160.

31. John Komlos, *Nutrition and Economic Development in the Eighteenth-Century Hapsburg Monarchy: An Anthropometric History* (Princeton, N.J.: Princeton University Press, 1989); Roderick Floud, Kenneth Wachter, and Annabel Gregory, *Height, Health and History: Nutritional Status in the United Kingdom, 1750–1980* (Cambridge: Cambridge University Press, 1990). For a critique of this method, see James C. Riley, "Height, Nutrition, and Mortality Risk Reconsidered," *Journal of Interdisciplinary History* 24 (1994): 465–71.

32. Thomas McKeown, in *The Modern Rise of Population* (New York: Academic Press, 1976), argued forcefully that nutritional level (not improvements in medicine) was the chief determinant of population growth in Europe. The nutritional thesis has been challenged by Massimo Livi-Bacci, *Population and Nutrition: An Essay on European Demographic History,* trans. Tania Croft-Murray (Cambridge: Cambridge University Press, 1991).

33. Stearns, *European History,* 15.

34. For spectators at feasts, see Per Bjurstrom, *Feast and Theatre in Queen Christina's Rome*, Nationalmusei skriftseries, no. 14 (Stockholm, 1966), 52–58; Peter Burke, *The Fabrication of Louis XIV* (New Haven, Conn.: Yale University Press, 1992), 87; Barbara Wheaton, *Savoring the Past: The French Kitchen and Table from 1300–1789* (Philadelphia: University of Pennsylvania Press, 1983), 134–35.

35. André Félibien, *Les plaisirs de l'isle enchanté*, cited in Wheaton, *Savoring the Past*, 135.

36. Jeffrey M. Pilcher, *¡Que vivan los tamales!* (Albuquerque: University of New Mexico Press, 1998), 105.

37. Redcliffe Salaman, *The History and Social Influence of the Potato* (1949; reprint, Cambridge: Cambridge University Press, 1970), chaps. 11–19; Arturo Warman, *La historia de un bastardo: Maíz y capitalismo* (Mexico: Fondo de Cultura Económica, 1988), chaps. 6, 7, 10, and 11; Sucheta Mazumdar, "The Impact of New World Crops on the Diet and Economy of India and China, 1600–1900," unpublished paper for a conference titled "Food in Global History," University of Michigan, 1996.

38. For England, see John Burnett, *Plenty and Want: A Social History of Diet in England from 1815 to the Present Day* (1966; reprint, London: Methuen, 1983), part 2; for France, *Theodore Zeldin, France 1848–1945*, vol. 2 (Oxford: Clarendon, 1974), 725–30; for Germany, *Hans Teuteberg und Gunter Wiegelmann, der Wandel der Nahrungsgewohnheiten unter dem Einfluss der Industrialisierung* (Göttingen: Vandenhoeck & Ruprecht, 1972).

39. Benjamin Disraeli, *Sybil, or The Two Nations, Book 2* (London: H. Colbourn, 1845), chap. 5.

40. Burnett, *Plenty and Want*, 128, 133; and Harvey Levenstein, *Revolution at the Table: The Transformation of the American Diet* (Oxford: Oxford University Press, 1988), 31–32.

41. Christian Petersen, *Bread and the British Economy, ca. 1770–1870*, ed. Andrew Jenkins (Aldershot, U.K.: Scolar Press, 1995), chaps. 2 and 4; Edward Thompson, "The Moral Economy of the English Crowd in the Eighteenth Century," *Past and Present* 50 (1971): 76–136, esp. 81.

42. Artusi Pellegrino, *The Art of Eating Well*, trans. Kyle M. Phillips II (New York: Random House, 1996), 76–77.

43. Emiko Ohnuki-Tierney, *Rice as Self: Japanese Identities through Time* (Princeton, N.J.: Princeton University Press, 1993), 41; Camellia Panjabi, *50 Great Curries of India* (London: Kyle Cathie, 1994), 185.

44. Camellia Panjabi, "The Non-emergence of the Regional Foods of India," in *Disappearing Foods: Proceedings of the Oxford Symposium on Food and Cookery*, ed. Harlan Walker (Totnes, U.K.: Prospect Books, 1995), 146–49; Owen, *Indonesian Regional Cooking*, introduction; Rachel Laudan, *The Food of Paradise: Exploring Hawaii's Culinary Heritage* (Honolulu: University of Hawaii Press, 1996), 7–8, 209–10.

45. This piece had its origins in a paper I gave at a conference at Oregon State University in 2000. It was published in the first issue of *Gastronomica* in 2001. Since then it has been reprinted multiple times, most recently in 2015 in the online magazine *Jacobin*, where it has received over twenty-five thousand hits.

46. Rachel Laudan, *Cuisine and Empire: Cooking in World History* (Berkeley: University of California Press, 2013), 42–55, 215–55, 280–90.

47. For republican cuisine in Antiquity, see Laudan, *Cuisine and Empire*, 74–79; in the Dutch Republic, Simon Schama, *The Embarrassment of Riches: An Interpretation of Dutch Culture in the Golden Age* (New York: Knopf, 1987), chap. 3, and Laudan, *Cuisine and*

Empire, 225–33; and in the modern period, Schama, *The Embarrassment of Riches*, 252–77, 313–23, and James E. McWilliams, *A Revolution in Eating: How the Quest for Food Shaped America* (New York: Columbia University Press, 2005), chap. 8.

48. Warren James Belasco, *Appetite for Change: How the Counterculture Took on the Food Industry, 1966–1988* (New York: Pantheon Books, 1989).

49. Carlo Petrini, *Slow Food: The Case for Taste*, trans. William McCuaig (New York: Columbia University Press, 2004); Lesley Chamberlain, "Rousseau's Philosophy of Food," *Petits Propos Culinaires* 21 (1985): 9–16; Laudan, *Cuisine and Empire*, 222–23; Michael Pollan, *The Omnivore's Dilemma: A Natural History of Four Meals* (New York: Penguin Press, 2006), 391–411.

50. "Doing Well by Eating Well," *Atlantic Monthly*, March 1999, https://www.theatlantic.com/magazine/archive/1999/03/doing-well-by-eating-well/377485/.

51. Jane Levi, "Charles Fourier versus the Gastronomes: The Contested Ground of Early Nineteenth-Century Consumption and Taste," *Utopian Studies* 26, no. 1 (2015): 41–45; Laudan, *Cuisine and Empire*, 313–23.

52. The inspiration for the "system" in which producers and consumers are exploited by capitalists came from Immanuel Wallerstein, *The Modern World-System 1* (New York: Academic Press, 1976). It is most clearly worked out for the food by Raj Patel, *Stuffed and Starved: From Farm to Fork, the Hidden Battle of the World Food System* (London: Portobello Books, 2007).

Contributors

KEN ALBALA is professor of history at the University of the Pacific in Stockton, California. He has authored or edited twenty-five books including academic monographs, cookbooks, encyclopedias, and other reference works. He was coeditor of the journal *Food Culture and Society* and edited a food series for Rowman and Littlefield including about fifty titles. His latest book is *Noodle Soup: Recipes, Techniques, Obsession.* His course *Food: A Cultural Culinary History* is available from The Great Courses and free as a podcast. He is currently walking with wine.

AMY BENTLEY is a professor in the Department of Nutrition and Food Studies at New York University. A historian with interests in the social, historical, and cultural contexts of food, she is the author of *Inventing Baby Food: Taste, Health, and the Industrialization of the American Diet* (2014) and *Eating for Victory: Food Rationing and the Politics of Domesticity* (1998) and the editor of *A Cultural History of Food in the Modern Era* (2011). Current projects include a history of food in U.S. hospitals, and a study of food production in cloistered religious communities. She serves as editor of *Food, Culture, and Society: An International Journal of Multidisciplinary Research.*

CHARLOTTE BILTEKOFF is an associate professor of American studies and food science and technology at the University of California, Davis. She teaches courses in food studies and innovation while building bridges between scientific and cultural approaches to food and health. She is the author of *Eating Right in America: The Cultural Politics of Food and Health* (2013). Current research includes a book project on the politics of expertise in contemporary debates about "processed food" and a National Science Foundation–funded collaboration investigating the California agro-food tech sector.

MATTHEW MORSE BOOKER is associate professor of American environmental history at North Carolina State University, where he directs the Science, Technology & Society program, coordinates the interdisciplinary Visual Narrative initiative, directs an oral history archive on agricultural genetic engineering, teaches agricultural and urban history, and mentors graduate students in six departments. His book *Down by the Bay: San Francisco's History between the Tides* (2013) detailed the transformation of that urban estuary from food producer to real estate. His current research explores the rise and fall of American oysters as a staple food of early twentieth-century industrial cities.

PETER A. COCLANIS is Albert R. Newsome Distinguished Professor of History and director of the Global Research Institute at UNC–Chapel Hill. He works primarily on the economic history of the United States and of Southeast Asia. He is past president of the Agricultural History Society and an elected Fellow of the society. He has written a great deal over the years on rice—its production and trade—all over the world. He is completing a global history of rice and is coeditor of *Rice: Global Networks and New Histories* (2015), which was named an "Outstanding Academic" book by *Choice*.

TRACEY DEUTSCH is associate professor of history at the University of Minnesota. Her areas of focus include modern U.S. history, critical food studies, gender and women's history, and the history of capitalism. Her published work includes *Building a Housewife's Paradise: Gender, Government, and American Grocery Stores, 1919–1968* and numerous essays on food, labor, and the gendered politics of consumption. From 2016 to 2019, she was also the Imagine Chair of Arts, Design and Humanities. She is currently researching Julia Child and mid-twentieth-century domesticity.

S. MARGOT FINN is a lecturer in applied liberal arts at the University of Michigan, where she teaches classes on food, obesity, and liberal education. Her first book is *Discriminating Taste: How Class Anxiety Created the American Food Revolution* (2017), which won the 2018 prize for best first book from the Association for the Study of Food and Society.

RACHEL LAUDAN is senior research fellow in the Institute for Historical Studies at the University of Texas at Austin. Following a first career as a historian of science, she turned to the history of food. Author of *Cuisine and Empire: Cooking in World History* (2013), in 2018 she received the annual Paradigm Award of the Progressive Institute "in recognition of the progressive story she tells about food and farming." She blogs at www.rachellaudan.com.

CHARLES C. LUDINGTON is teaching associate professor of history at North Carolina State University, where he teaches classes in British, Irish, Continental European, and food history. His published work includes *The Politics of Wine in Britain: A New Cultural History* (2013). From 2015 to 2017 he was a Marie Sklodowska-Curie Senior Research Fellow at University College Cork and Université de Bordeaux–Michel Montaigne. He is currently writing a book about the role of Irish merchants in Bordeaux in the eighteenth century, including their contributions to the development of the Bordeaux wine trade.

SARAH LUDINGTON is associate dean of academic affairs and professor at Campbell University Law School. She teaches and writes about constitutional law, civil and administrative procedure, information privacy, and the First Amendment.

MARGARET MELLON is currently a consultant on biotechnology and sustainable agriculture. She founded the Food and Environment Program at the Union of Concerned Scientists in 1987. She served three terms on the U.S. Department of Agriculture's Advisory Committee on Biotechnology and 21st Century Agriculture and was named a Fellow of the American Association for the Advancement of Science (1994). Her published works include *Ecological Risks of Engineered Crops* (1996) and *Hogging It! Estimates of Antimicrobial Abuse in Livestock* (2001). She holds a doctorate in molecular biology and a law degree from the University of Virginia.

STEVE STRIFFLER is the director of the Labor Resource Center and professor of anthropology at the University of Massachusetts, Boston. His first book, *In the Shadows of State and Capital* (2002), examined labor conflicts in the banana industry in Ecuador. His second book, *Chicken: The Dangerous Transformation of America's Favorite Food* (2005), explored labor and migration in relation to the poultry industry. His most recent book is *Solidarity: Latin America and the U.S. Left in the Era of Human Rights* (2019).

ROBERT T. VALGENTI is professor of philosophy and chair of the Department of Religion and Philosophy at Lebanon Valley College. His research and teaching covers contemporary Italian philosophy, hermeneutics, biopolitics, and the philosophy of food. He is the director and founder of EAT (engage, analyze, transform), an undergraduate research group at Lebanon Valley College whose interventions aim to improve the ethical, environmental, cultural, and nutritional profile of the college dining experience. He is also a member of the Menus of Change University Research Collaborative and a desk editor for the journal *Gastronomica*.

Index

MIX
Paper from
responsible sources
FSC® C008955
FSC
www.fsc.org